CONSTRUCTING NUMBERS

An Inquiry-Based Capstone Mathematics Course

First Edition

By Mark Daniels and Efraim Armendariz

University of Texas - Austin

cognella®

academic publishing

Bassim Hamadeh, CEO and Publisher
Michael Simpson, Vice President of Acquisitions
Jamie Giganti, Managing Editor
Jess Busch, Senior Graphic Designer
Marissa Applegate, Acquisitions Editor
Gem Rabanera, Project Editor
Alexa Lucido, Licensing Coordinator

First published in the United States of America in 2015 by Cognella, Inc.

Cover images:
Copyright © 2012 by Depositphotos / siminitzki.
Copyright © 2010 by Depositphotos / Karuka.

Printed in the United States of America

ISBN: 978-1-63189-459-6 (pbk) / 978-1-63189-460-2 (br)

cognella®
academic publishing

www.cognella.com   800-200-3908

Contents

Preface

Constructing Numbers: A Creative and Connected Approach

About This Text

The presentation of topics in this text is done in a "discovery-based" format, so as to invite the learner/reader to actively participate in the exercises and proofs that develop the material in a logical manner. We also include explorations that serve as activities for enrichment that are not part of the development of the material.

This textbook is intended to be used primarily for a one-semester, upper-division mathematics course. The topics covered in this text serve to synthesize material from elementary analysis, linear algebra, algebraic structures, and geometry, while providing a historical perspective of how mathematical thought develops. We believe that the material and mathematical connections made in this text are vital and worthwhile for all university undergraduate mathematics students to experience.

This book is also extremely well suited for a discovery-based mathematics course for preservice secondary teachers. The course topics were born out of notes used for a capstone mathematics course that is taught to upper-division mathematics majors in the UTeach Program in Natural Sciences at the University of Texas. UTeach mathematics majors, in addition to earning an undergraduate degree in mathematics, are seeking state teaching certification in middle school or high school mathematics.

Suggestions for Using This Text

Instructors of a course using this text are encouraged to present the material of the book in an inquiry-based method. That is, allow students to work through the exercises and theorems of the text collaboratively and devote considerable class time to student presentation of results. The instructor is further encouraged to take on the role of facilitator in the course, in that one of the goals of the course would be to bring students to the point where students would not simply present results but also challenge and correct each other concerning the formal logic and methods employed in their presentations. This is a book and a course about getting students to think deductively and logically about fundamental ideas in mathematics.

Further, instructors are encouraged to add any relevant topics, exercises, and theorems to those presented in the text that seem necessary based on student makeup and individual goals for the course. The authors welcome any suggestions along those lines. Please feel free to use the material of the text as a springboard for exploration of tangential topics and connections based on student interest and discussion.

Last, to the student, we encourage you to view the topics, exercises, and explorations of the text as vehicles to be used toward thinking deeply about concepts and connections between concepts that you may have seen before but not in the same depth or context. The book also allows you to experience the axiomatic development of important mathematics foundational topics. We believe that this material is essential for those undergraduate mathematics majors who are interested in pursuing graduate mathematics studies. If you are a mathematics major who is also interested in teaching mathematics, this text will present the theory and logic necessary for you to "fill in" the number line and then the Cartesian plane. This will form the foundation for many of the topics in mathematics, such as those related to functions and function properties that you will teach to your future students.

A Note to Instructors

Constructing Numbers: A Creative and Connected Approach

We, the authors, welcome you to our text. We would like to provide you, as an instructor for the course associated with this text, with some background information as to the intended target audience for *Constructing Numbers* and the authors' intentions in creating the book. We assume that students in a course that uses *Constructing Numbers* are mathematics majors of junior- or senior-level standing. The expectation is that these students have completed Calculus and Linear Algebra plus two proof-based courses, such as Discrete Mathematics, Number Theory, Elementary Analysis, Modern Geometry, or Elementary Algebraic Structures. We do not consider the text to be a primary source for a methods course in mathematics education, although we believe that it can be used as an excellent supplementary text for such a course. Outside of the preservice education realm, we consider *Constructing Numbers* to be an appropriate text for a capstone course in undergraduate mathematics.

The overall goal of *Constructing Numbers* is to have students "do" mathematics—that is, to have students engage in thinking about the connectedness that exists between various basic areas of mathematics. In addition, students should work to provide rigorous arguments at different levels that support the development of these connections. Our hope is that, as a result of working through *Constructing Numbers*, students will more deeply understand the discipline of mathematics and the fact that if one does not ask "why" when engaging in "doing" mathematics, the processes experienced are strictly mechanical.

We have tried to provide a unique, well-designed, and focused collection of essential mathematics topics in *Constructing Numbers*. For this reason, we have purposely not stressed the typical "real-life" applications associated with the theory and topics presented in the text. It must be noted, however, that the mathematics presented in this text was developed in response to real-world needs. We feel that the goal of a course based on our text should be to enable students to "go beyond" the specific applications associated with the material of the text. By this we mean that the thinking needed to problem solve and the presentation of logical arguments are skills that can be gained or enhanced in students as a result of working through *Constructing Numbers*. These are skills that can be applied generally to "real-world" needs. This is reinforced by the intention of the authors that the book be used in a discovery or open-forum environment, where students present logical arguments and are critiqued by peers, with the instructor acting as a facilitator. This is our perception of what we mean by any of the terms that might be applied to an "open-forum" method of instruction. These terms include "discovery," "Moore method," "modified Moore method," "inquiry based," "Socratic," and "problem solving." We are not concerned with the fine distinctions but rather the commonality of experience present in such methodology.

We have also taken great care to choose topics covered in the text carefully. The last two chapters related to probability and group theory were added at the suggestion of respected colleagues. The authors firmly believe that the inclusion of these extra topics in *Constructing Numbers* enhances the text as a connected and coherent endeavor. However, it is important to remember that the topics of this text should not be covered by instructors with the mind-set of presenting the material in a "race-to-the-finish" mode. It is paramount that students in the course be allowed the time to struggle with the topics and think deeply about connections made between topics in the course. Cover what you can well, but don't worry if you don't make it to the end of the text.

Introduction

A Brief History of Numbers

Numbers and counting are essential elements of all cultures. The development of numbers and number systems began as a way to fill the utilitarian need of being able to count discrete objects, such as livestock, and to record these numbers, which were used to signify a quantity. The seemingly simple leap from counting quantities to recording quantities required ancient cultures to devise ways to represent numbers using symbols. This was accomplished in many ways by many different cultures.

Different cultures devised varying systems using diverse symbols to fulfill the functional need to count and quantify. Certain native Aboriginal tribes of Australia used a very simple numbering system consisting of 0, 1, and many. It is well known, for example, that the Babylonians (circa 2000 BC) used 60 as a base unit for their numbering system. Remnants of this system survive today. The base value for our timekeeping system of 60 seconds and the $360°$ of arc length in a circle had their origins in the Babylonian system. The Babylonians are also credited with devising a place-value decimal notation. The Egyptians developed a novel way of representing rational number values of the form a/b as sums of *unit fractions* of the form $1/m + 1/n +$ ..., where $m, n, ...$ are integers. The more recognizable Hindu-Arabic system of writing numerals as 1, 2, 3, ..., 9 and an assigned value of 0 started to take hold around 1200 AD.

Along the way, operations such as addition and multiplication were developed and used with numbers, and discrete systems gave way to continuous systems as the number line was "filled in." Early "trial-and-error" methods fell by the wayside, as did generalizations that were based on a limited number of specific observations or experiments. The processes that led to our modern and uniform system of numbers took hundreds of years to develop as logic, theory, and abstraction were infused into mathematical thinking. Definitions, axioms, and formal verifiable statements led to the consistent and systematic development of mathematics as a subject of study in its own right.

While this book is not about the history of the development of number systems, it does strive to allow the active reader to axiomatically experience the development of the field operations that pertain to "numbers," the filling in of the real number line, and the constructibility of numbers in the plane.

1.

Numbers as Axiomatic Systems

The Journey Begins

Much have I travel'd in the realms of gold,
And many goodly states and kingdoms seen;
Round many western islands have I been
Which bards in fealty to Apollo hold.
Oft of one wide expanse had I been told
That deep-brow'd Homer ruled as his demesne;
Yet did I never breathe its pure serene
Till I heard Chapman speak out loud and bold:
Then felt I like some watcher of the skies
When a new planet swims into his ken;
Or like stout Cortez when with eagle eyes
He star'd at the Pacific—and all his men
Look'd at each other with a wild surmise—
Silent, upon a peak in Darien.

"On First Looking Into Chapman's Homer"
by John Keats

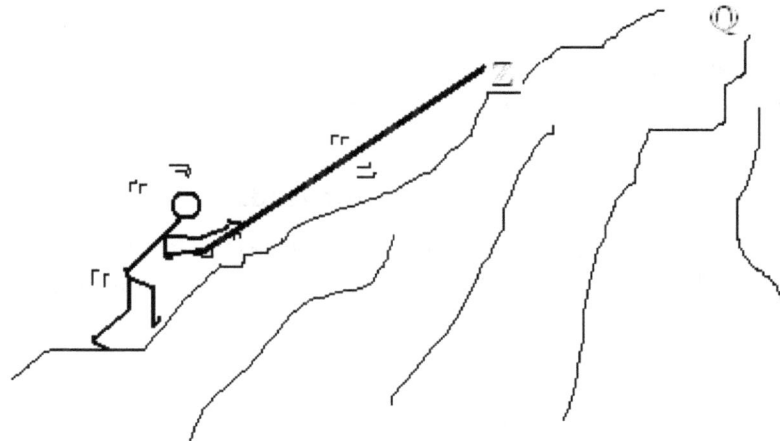

W e will be considering sets made up of objects. We will always assume that a set has at least one object; that is, the word *set* will mean a non-empty set.

The sets that we consider will have special properties, and we will call such a set a *field of numbers*, or simply a *field*. The objects belonging to the set will be called *numbers*. Initially, we do not ascribe any concrete representations to these number systems. However, the reader may wish to keep in mind that the number systems of arithmetic described in the introductory chapter are models for the concepts introduced.

Our number systems have certain basic properties in common. The reasons for identifying a basic collection of properties that should be held in common are threefold.

The first reason is for purposes of identification. We can use the basic properties to determine whether a set falls into the category of fields. Thus, if we can verify that a set has all the prescribed basic properties, we are dealing with a number system called a field. Second, these properties serve as rules for operating with the objects in a field. Finally, use of the rules permits us to deduce, using logical reasoning, other properties that are present in our system.

The basic properties are called *axioms*. An *axiom* is a foundational statement that is assumed to be true and requires no verification. A consequence of the axioms, that is, an additional property that is deduced logically from the axioms, is referred to as a *theorem*. Theorems are usually presented as statements written in an "if A, then B" format. An equivalent format for the statement of a theorem is "assume A, then B." A theorem might also be a statement in the form of an assertion that "A is B."

Theorems are logical consequences of the axioms or other previously deduced theorems. Although statements will be made in the sequel that are identified as theorems, what will not be present are the logical steps that lead to the statement of the theorem. Therefore, in the spirit of "discovery-based" learning or inquiry, the learner/reader is expected to obtain all steps leading to the proof of a theorem. Exercises are also included to assist the learner/reader in developing familiarity with the material as well as a fuller understanding of the subject matter. Furthermore, results established in exercises can be used in proving subsequent theorems.

We will make the additional assumption that we have at hand the set of integers with their addition and multiplication. Last, the set of integers will be denoted by the letter $\mathbb{Z}$. Thus, $\mathbb{Z}$ consists of the 0, the counting numbers $\{1, 2, 3, \ldots\}$, and their "minuses" $\{-1, -2, -3, \ldots\}$.

The Field Axioms

We begin with a set of objects; we will designate the set by the symbol S and the objects in the set S by lowercase letters, such as x, y, r, s, and so on. The eleven axioms that serve to identify a *field* now follow.

Axioms for Addition

A1.   There is a well-defined operation "+" on S, which associates for any two objects x and y in S another object in S, called the sum of x and y, and denoted by the expression $x + y$.

A2.   For any two objects x and y in S, $x + y$ is the same as $y + x$.

A3.   For any three objects x, y, and z in S,
$(x + y) + z$ is the same as $x + (y + z)$.

A4.   The set S contains an object o that has the following property:
for any object x in S, $x + o$ is the same as x.

A5. For any object x in S, there is an object x^* with the following property: $x + x^*$ is the same as the object o identified in A4.

Axioms for Multiplication

M1. There is a well-defined operation "$\cdot$" on S, which associates for any two objects x and y in S another object in S, called the product of x and y, and denoted by the expression $x \cdot y$.

M2. For any two objects x and y in S, $x \cdot y$ is the same as $y \cdot x$.

M3. For any three objects x, y, and z in S, $(x \cdot y) \cdot z$ is the same as $x \cdot (y \cdot z)$.

M4. The set S contains an object u, distinct from o in A4, that has the following property: for any object x in S, $x \cdot u$ is the same as x.

M5. For any object x in S different from the object o identified in A4, there is an object $x^{\#}$ with the following property: $x \cdot x^{\#}$ is the same as the object u identified in M4.

Axiom Connecting Addition and Multiplication

D. For any three objects x, y, and z in S, $x \cdot (y + z)$ is the same as $x \cdot y + x \cdot z$.

A set S of objects satisfying properties A1 through A5, M1 through M5, and D will be referred to as a *field of numbers*, or, simply, *a field*, and the objects in S will be called *numbers*. The properties of addition and multiplication are also commonly stated as follows: addition in S, as well as multiplication in S, is commutative (A2 and M2) and associative (A3 and M3), and multiplication is distributive over addition (D).

Standing on a Firm Foundation

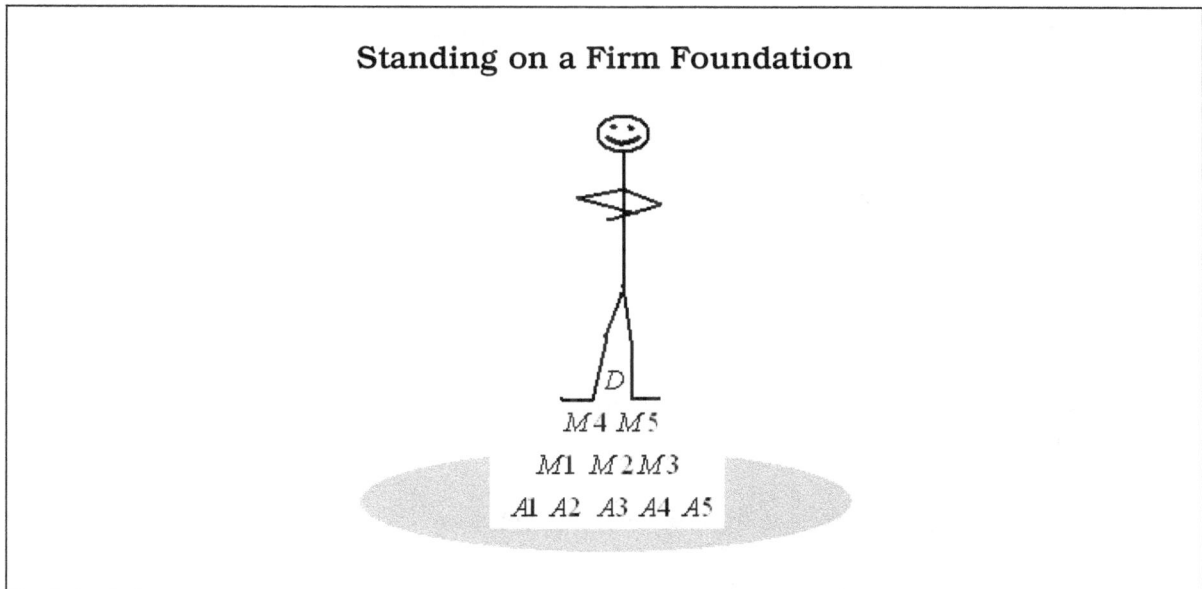

Equality

Given a number system S, we will write "$x = y$" to indicate that x is the same number as y. Thus, for example, A2, can be restated as follows: for any two numbers x and y in S, $x + y = y + x$.

Exercise Set 1

> Restate the axioms using "=."

We will use the properties of equality for numbers x, y, and z in a number system:

E1: $x = x$
E2: If $x = y$, then $y = x$.
E3: If $x = y$ and $y = z$, then $x = z$.
E4: If $x = y$, then $x + z = y + z$
E5: If $x = y$, then $z \cdot x = z \cdot y$.

Basic Consequences of the Axioms

You will now be asked to construct logical arguments using the axioms and properties of equality to verify the assertion made in the statements of theorems. Here is our first statement.

THEOREM 1

> The object o identified in axiom 4 is unique. That is, if a is a number with the property that $x + a = x$ for all x in S, then $a = o$.

Once the uniqueness of a special number has been shown, it is possible to assign a special symbol to designate the number. For example, the number o identified in axiom 4 will now be designated by 0 (and given the name "zero"). The next few exercises establish the uniqueness of other objects identified in the field axioms.

Exercise Set 2

Let S be a field. Verify each of the following statements:

1. The number u identified in axiom M4 is unique; it will now be designated by the symbol 1.

2. For each x in S, there is only one number associated with x having the property stated in axiom A5; it will now be designated by the symbol $-x$.

3. For each x in S different from 0, there is only one number associated with x having the property stated in axiom M5; it will now be designated by the symbol x^{-1}.

We need to emphasize the results established in theorem 1 and in exercise set 2. If we are working within a field,

a. The symbol 0 is the unique number in S with the property that $x + 0 = x$ for all x in S (note that $0 + x$ is also equal to x for all x in S); 0 is called the *additive identity of S*.

b. The symbol 1 is the unique number in S with the property that $1 \cdot x = x$ for all x in S (note that 1 is different from 0 and that $x \cdot 1 = x$ for all x in S); 1 is called the *multiplicative identity of S*.

c. For each number x in S, $-x$ represents the unique number in S, which when added to x yields 0; $-x$ is called the *additive inverse of x*.

d. For each x in S different from 0, x^{-1} represents the unique number in S, which when multiplied with x yields 1; x^{-1} is called the *multiplicative inverse of x*.

We will find it helpful to have additional properties that assist in carrying out calculations within a field. The next results provide such properties.

THEOREM 2

Let S be a field. For any $x \in S, 0 \cdot x = 0$.

THEOREM 3 (PROPERTIES OF ADDITIVE INVERSES)

Let S be a field. For any x, y in S, the following statements are true:

a. $-(x + y) = (-x) + (-y)$
b. $-(-x) = x$
c. $(-x)(-y) = xy$
d. $(-1)x = -x$

THEOREM 4

Let S be a field. For x, y in S, both different from 0,

a. $(xy)^{-1} = (x^{-1})(y^{-1})$
b. $(x^{-1})^{-1} = x$
c. $(-1)x^{-1} = (-x^{-1}) = -(x^{-1})$
d. $(-1)^{-1} = -1$

EXPLORATION 1—FUNCTIONS AS NUMBERS

1. Let F be the set of all real-valued continuous functions whose domain contains $[0, 1]$. For example, $f(x) = x^2$.
   Considering these functions as "numbers," determine which of axioms A1 through A5, M1 through M5, and D are satisfied. (Note: If f and g are in F, what is meant by $f + g$ and $f \cdot g$?)

2. Explain why it is that

$$\frac{4+x^6}{x^3} = \frac{4}{x^3} + \frac{x^6}{x^3},$$

but

$$\frac{x^3}{4+x^6} \neq \frac{x^3}{4} + \frac{x^3}{x^6} ?$$

A Historical Side Trip on Fields

The term *field* is a rough approximation to the German word *koerper* and the French word *corps*, used in those languages to describe the number systems under discussion. European mathematicians formalized the concept in the mid-nineteenth century, recognizing that fields were the proper setting for discussing solutions of polynomial equations.

The solution of polynomial equations is a basic problem in mathematics and has a rich history. Babylonian (circa 400 BC), Greek (circa 300 BC), Hindu (circa 600 AD), and Arab (circa 800 AD) societies all appear to have acquired methods for solving linear and quadratic equations. We should note that these equations did not necessarily appear in the form in which they are presently studied, making use of a variable and coefficients. The description of the problem and the corresponding method used for solving the equation necessitated lengthy description.

One can imagine a problem posed in the following manner: find a number that when doubled and augmented by five units yields eleven units. Its solution might be given by the following statement: after reducing eleven units by five units, halve the result to obtain the desired number. A quadratic equation would be even more involved, as would the method for solving the equation. There appeared to be no unified method for describing a general linear equation $ax + b = 0$, for example, or a general quadratic equation $ax^2 + bx + c = 0$.

However, the Hindu mathematician Brahmagupta, in his treatise *Brahmasphuta siddhanta* on the physical world, written around the year 625 AD, did include a poetic description for solving quadratics that verbally described the quadratic formula. The general symbolic formula used in algebra today to solve a quadratic equation did not appear in Europe until the twelfth century, when the book *Liber embadorum* was published in 1145 by Abraham bar Hiyya Ha-Nasi, under the Latin name Savasorda.

The Italian mathematician Cardan is generally credited with determining an algorithm for solving cubic equations, and his method of substitution was readily adapted to use the quadratic formula to solve quartic equations. It was not until 1826 that N. H. Abel proved that there was no general procedure involving radicals (square roots, cube roots, etc.) that could be applied to any quintic equation. The result is also credited to Ruffini, because he had proposed a flawed argument that was very similar to that of Abel. The recognition that no algorithm existed for quintics was a gigantic step in the theory of equations and led to the formal development of an algebraic theory of fields that dealt with the existence of solutions of polynomial equations. The principal actor in this development was E. Galois, who extended the Abel-Ruffini result by showing the impossibility of using radicals to solve general nth-degree equations for all $n > 4$. The theoretical area of mathematics that deals with solutions of polynomial equations is now known as Galois theory.

EXERCISE SET 3

Describe verbally, without using mathematical symbols, a quadratic equation and how to use the quadratic formula to solve the quadratic equation.

EXERCISE SET 4

Let S be a field.

1. Determine the solution in S to the equation $ax + b = 0$ for $a, b \in S$ with $a \neq 0$.
2. If $a, b, c \in S$ and $ab = 0$, show that either $a = 0$ or $b = 0$.
3. Determine all solution(s) in S to the equation $x^2 = x$.
4. If $a, b \in S$ with $a \neq 0$, find a condition that ensures $ax^2 + bx + c = 0$ has a solution in S.
5. Let S be a number system, $T \subseteq S$, such that T is a number system relative to the operations defined on S. Show that both 0 and 1 are in T.

Infinite and Finite Sets

With the agreement that the non-empty set S is a set with at least one element, a set S is *infinite* if there is a one-to-one correspondence between S and a proper subset of S. A non-empty set S is *finite* if it is not infinite.

EXERCISE SET 5

1. Let $S = \{0, u\}$. Show that S is a finite set.
2. Let $S = \{0, u\}$. Is it possible to define "addition" and "multiplication" operations on S so that S is a number system? Tables such as

+	0	u
0		
u		

·	0	u
0		
u		

   may be useful. (For example, using the "+" table, one can define the sum of any two objects in S. Thus, the entry in the first row, second column should represent $0 + u$ and is either 0 or u.)

3. Repeat exercise 2 with the symbols $\{0, u, x\}$.
4. Repeat exercise 2 with 4 symbols $\{0, u, x, y\}$.
5. Repeat exercise 2 with 5 symbols $\{0, u, x, y, z\}$.
6. Repeat exercise 2 with 6 symbols $\{0, u, x, y, z, w\}$.
7. Based on these exercises, can you make a reasonable conjecture?

Special Sets of Numbers and Order

At this point it is useful to examine special examples of numbers, such as the set of *integers*. Recall that the set of integers will be denoted by the symbol $\mathbb{Z}$. Further, $\mathbb{Z}$ consists of the counting numbers $\{1, 2, 3, \ldots\}$, their *additive inverses* $\{-1, -2, -3, \ldots\}$, and 0.

Recall also that we assume the existence of $\mathbb{Z}$, together with the operations of addition and multiplication. We observe that axioms A1 through A5, D and M1 through M4 are satisfied but not M5. Thus, $\mathbb{Z}$ is not a field, although it comes close to being one.

The set $\mathbb{Z}$ also carries with it an *ordering*. We refer to the set $\{1, 2, 3, \ldots\}$ as the *positive integers* and the set $\{-1, -2, -3, \ldots\}$ as the *negative integers*, while 0 is neither a positive integer nor a negative integer.

A second basic set of numbers is the set of *rational* numbers, denoted by the symbol $\mathbb{Q}$. Here $\mathbb{Q}$ consists of all fractions a/b, where $a, b \in \mathbb{Z}$ and $b \neq 0$.

We recall that two fractions a/b and c/d are equal, $a/b = c/d$, if and only if $ad = bc$.

Further, we have addition and multiplication defined in $\mathbb{Q}$; specifically,

$$\frac{a}{b} + \frac{c}{d} = \frac{ad + bc}{bd} \text{ and } \frac{a}{b} \cdot \frac{c}{d} = \frac{ac}{bd}.$$

Observe that we can express each integer a as a fraction $a/1$; thus, $\mathbb{Q}$ contains all integers. Note also that all of axioms A1 through A5, M1 through M5, and D are satisfied in $\mathbb{Q}$, so $\mathbb{Q}$ is a field.

It is also true that $\mathbb{Q}$ can be partitioned into the three disjoint subsets consisting of $\{0\}$, the set of positive rational numbers, and the set of negative rational numbers, in such a way that all positive integers are positive rational numbers and all negative integers are negative rational numbers.

Using these two examples as models, we next introduce the concept of an *ordered field*.

Definition: Assume that S is a field. Then S is an *ordered field* if S can be partitioned into three disjoints sets P, $-P = \{-x \mid x \text{ is in } P\}$, and $\{0\}$, such that if

$$x, y \in P, \text{ then } x + y \in P \text{ and } x \cdot y \in P.$$

The elements of S belonging to P are called the *positive elements*, while those belonging to N are called the *negative elements*. Note that 0 is neither a positive element nor a negative element of S.

We next introduce the symbol ">" to represent order. We write $x > 0$, if and only if $x \in P$. If $x, y \in S$, we write

$$x > y \text{ if and only if } x - y \in P;$$

that is, $x > y$ if and only if $x - y > 0$.   (Note: "$x > y$" can be read as x is greater than y.)

THEOREM 5

Let S be an ordered field.

For any $x, y \in S$, exactly one of the following holds:

$$x = y, x > y, \text{ or } y > x.$$

i.   If $x, y \in S$ and $x > y$, then for any $z \in S$, $x + z > y + z$.

ii. If $x > y$ and $z > 0$, then $xz > yz$.

iii. If $x > y$ and $y > z$, then $x > z$.

EXERCISE SET 6

Let S be an ordered field. Show that

1. $1 > 0$

2. If $x > 0$, then $1/x > 0$.

3. If $x > 0$ and $y > x$, then $1/x > 1/y$.

4. Given x, y, a, and b in S with $y > x$ and $b > a$, then $y + b > x + a$.

5. If $y > x$, then there exists a in S, such that $0 < a$ and $x + a = y$.

6. $x + 1 > x$ for all x in S.

7. If $x > y$, then $-y > -x$.

8. How is a positive rational number defined by making use of the order present in $\mathbb{Z}$?

Considering notation, it is convenient to use the following convention:

$x < y$ if and only if $y > x$; furthermore,
$x \leq y$ if and only if $x = y$ or $x < y$.   (Note: That is, x is not greater than y.)

In preparation for dealing with the field of real numbers, which will be the subject of the next chapter, we introduce a concept, *absolute value*, that is used to measure the magnitude of numbers in an ordered field, as well as provide a measure of how far away two numbers in an ordered field are from each other.

Absolute Value

Definition: Let S be an ordered field. For a number x, we define the *absolute value* of x to be x if either $x = 0$ or if $x > 0$, and to be $-x$ if $x < 0$.

The absolute value of x is denoted by $|x|$. Thus,

$$|x| = \begin{cases} x, & \text{if } x \geq 0 \\ -x, & \text{if } x < 0 \end{cases}.$$

Exercise Set 7

Let S be an ordered field, and let $a \geq 0$. Show that

1. $|x| \leq a$ if and only if $-a \leq x \leq a$.

2. If x and y are numbers in S, then $|xy| = |x||y|$.

3. If x, y are numbers in S, then $|x + y| \leq |x| + |y|$.

Definition: For numbers x, y in an ordered field S, the *distance between x and y* is defined to be $|x - y|$, and is denoted by $d(x, y)$.

Exercise Set 8

Let S be an ordered field.

If x, y, z are numbers in S, show that

1. $d(x, y) \geq 0$ *and* $d(x, y) = d(y, x)$.

2. $d(x, y) = 0$ if and only if $x = y$.

3. $d(x, z) \leq d(x, y) + d(y, z)$.

2. The Real Numbers

As a result of introducing an order relation on $\mathbb{Z}$ and, therefore, on $\mathbb{Q}$, we are now in a position to obtain a geometric, or pictorial, representation for these sets of numbers. Keep in mind that alternate representations are possible. However, the one we use, the number line, is the one most commonly used. It is easy to describe and appears to be consistent with our intuition.

We choose a point in the plane and a line passing through the point. We assume that the line is parallel to the Earth's surface in our immediate surroundings (as opposed to a vertical line); we also assume that the line extends without bound in both directions away from the selected point.

The number 0 is assigned to the selected point. We next select a fixed length, L, and measure points to the right of 0 of lengths $L, 2L, 3L, \ldots$ This establishes a correspondence between the positive integers $\{1, 2, 3, \ldots\}$ and the points $L, 2L, 3L, \ldots$ on the line. The negative integers $\{-1, -2, -3, \ldots\}$ are assigned to the points obtained by measuring lengths $L, 2L, 3L, \ldots$ to the left of 0. With this correspondence we now identify positive integers with points on the line to the right of 0, while negative integers are identified with points on the line to the left of 0.

Addition and multiplication in $\mathbb{Z}$ translate to the points on the number line in the following manner: if x and y correspond to the points xL and yL, respectively, then $x + y$ and xy correspond to the points $(x + y)L$ and $(xy)L$, respectively.

The next step is to identify a rational number a/b with the point corresponding to the distance $(a/b)L$. As with integers, a rational number is positive if it corresponds to a point to the right of 0, and it is negative if it corresponds to a point to the left of 0.

We can now interpret what we mean geometrically by the relation "$x > y$." Specifically,

$$x > y \text{ if } x \text{ is to the right of } y.$$

Exercise Set 9

Verify that statements i through iii in theorem 5 are satisfied by the members of $\mathbb{Q}$ using the preceding interpretation of "$x > y$."

At this point, we are faced with the question of whether or not the assignment of rational numbers uses up all points on the number line. In the spirit of inquiry you might consider the question written as an exercise in the following form.

Exercise Set 10

1. Do there exist integers a, b with $b \neq 0$, such that $5 = \dfrac{a^2}{b^2}$, for example?
   (Then $5b^2 = a^2$.)

2. What does this imply about a relative to 5 and thus about b relative to 5?

3. What can be concluded if a and b are assumed to have no common divisor; that is, if the fraction $\dfrac{a}{b}$ is assumed to be in reduced form?

Theorem 6

If x is a number, such that $x^2 = 2$, then x is not a rational number.

Theorem 7

Use a geometric argument to show that there exists a positive number x, such that $x^2 = 2$.

Exercise Set 11

Based on exercise set 10 and theorem 6, can you state and prove a general result?

The previous two theorems suggest that there may be members of an ordered field of numbers that correspond to points on our selected line. Such a number system has to respect the addition, multiplication, and order existing in $\mathbb{Q}$.

Note: We will assume that such an ordered field exists, such that all points on the selected line correspond to members of the field. The assumed field is called the *field of real numbers* and is designated by the symbol $\mathbb{R}$. Thus, we are assuming that $\mathbb{R}$ is an ordered field containing the rational numbers, and we will proceed to describe an axiomatic system that justifies the identification of members of this field with the points on the line.

Observe that we will not show the existence of a field called the real numbers. To do so requires rather sophisticated mathematics. Historically, the non-geometric development of the real numbers (i.e., independent of a line) was carried out in the late nineteenth century and early twentieth century independently by R. Dedekind, G. Cantor, and A. Cauchy. However, we should note that even without a precise foundation for the real numbers, considerable and significant mathematics was carried out by mathematicians such as Newton (1642–1727), Leibniz (1646–1716), Euler (1707–1783), and Hindu mathematicians (circa 600 AD).

Analysis of the Real Line 1

Definition: Let H be a set of real numbers. The real number x is said to be an *upper bound* for H if $a \leq x$ for all $a \in H$.

Definition: The real number x is a *least upper bound (lub)* if x is an upper bound for H and $x \leq y$ for each upper bound y of H.

Theorem 8

If a set H has a least upper bound x, then x is unique.

Theorem 9

Suppose that x is an upper bound for the set H; then x is the least upper bound for H if and only if for each $\varepsilon > 0$ there exists $a \in H$, such that $x - \varepsilon < a$.

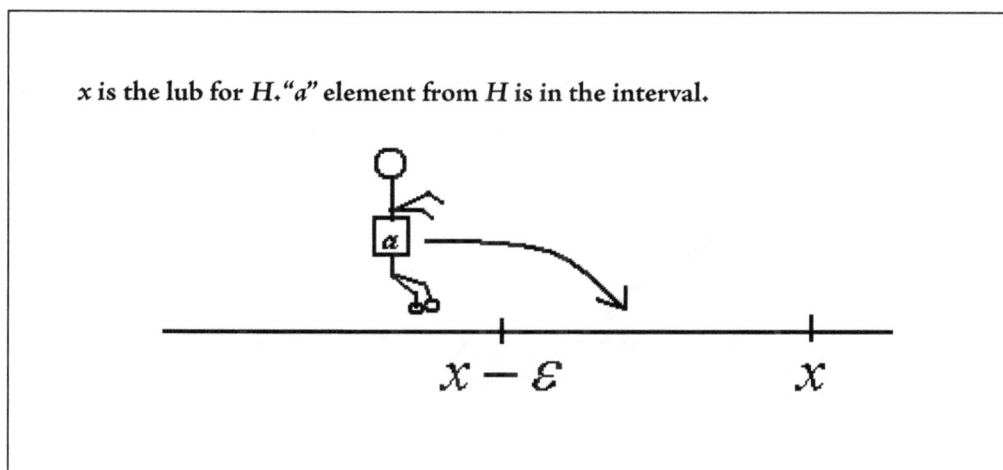

x is the lub for H. "a" element from H is in the interval.

We can now present a significant axiom that does two things. First, it distinguishes the rational numbers, in that the axiom fails to hold for this field. Second, it enables us to justify the identification of each of the points on the number line with a real number.

The completeness axiom for the real numbers (existence of the least upper bounds for bounded sets): Let H be a set of real numbers, and assume that there is a real number that is an upper bound for H. Then there is a real number that is a least upper bound for H.

Exercise Set 12

1. Provide definitions for the terms *lower bound* and *greatest lower bound*.

2. State the analogue of the completeness axiom for lower bounds.

3. Prove that the two statements are equivalent; that is, if one has the completeness axiom for upper bounds, then one has the completeness axiom for lower bounds, and vice-versa.

The Archimedean Property

The *Archimedean property* gives us insight concerning the comparative sizes of numbers that can be extremely large or extremely small (infinitesimals). Archimedes (circa 250 BC) seemed to be a master at using the concepts of the "large" and the "small" to his advantage. For example, Archimedes was able to approximate the value of π by considering a circle as a polygon containing a large number of sides, each of which was extremely small. Archimedes also finished the work of Aristotle by defining a relationship called "the law of the lever." This statement defines a basic principle of physics that describes the relationship between unequal forces applied to a lever about a fulcrum point at unequal distances. Archimedes is attributed with making the claim that, given a large enough lever and a place to stand, he could use a relatively small force to move the Earth.

Theorem 10 (THE ARCHIMEDEAN PROPERTY)

Given any two positive real numbers x, y, there is an integer n, such that $nx > y$.

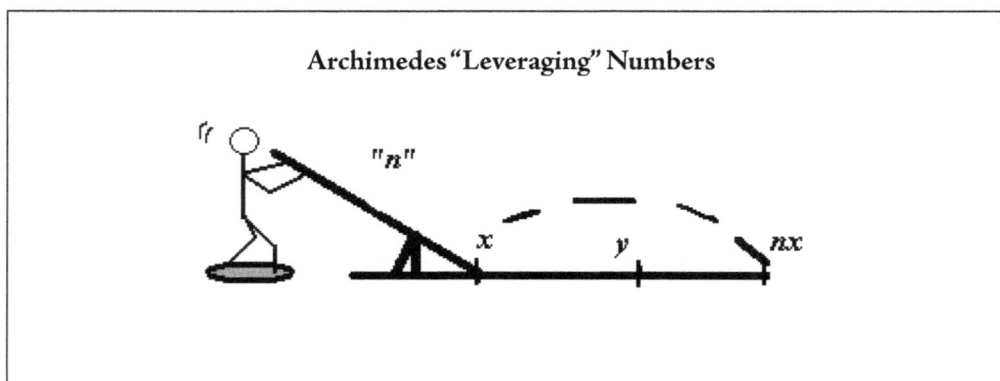

Archimedes "Leveraging" Numbers

Archimedes, in *The Quadrature of the Parabola*, credits Eudoxus as the inspiration for the "Archimedean postulate," which was written as follows: "When two spaces are unequal, it is possible to add to itself the difference by which the lesser is surpassed by the greater, so often that every finite space will be exceeded." Archimedes' work laid the foundation for infinitesimal calculus and the idea that a *complete field* satisfies *the postulate of continuity*; that is, least upper bounds exist for non-empty sets of real numbers.

EXPLORATION 2: USING "SMALL NUMBERS" TO YOUR ADVANTAGE

Prove these statements:

1. Let $x > 1$. Show that if k is a counting number, such that
$$k > 2x,$$

then (i) $\left(x - \dfrac{1}{k^4}\right)^2 > x^2 - \dfrac{1}{k}$ and

(ii) $\left(x + \dfrac{1}{k^4}\right)^2 < x^2 + \dfrac{1}{k}.$

2. If $a, b \in \mathbb{R}$, then $a^2 + b^2 \geq 2ab$. Also, state when equality holds for this statement.

Exercise Set 13

Prove the following

1. If $\sqrt{a}$ exists for any $a > 1$, then $\sqrt{a}$ exists for any $0 < a \leq 1$.

2. $\sqrt{2} \notin \mathbb{Q}$. Also show that, in general, $\sqrt{p} \notin \mathbb{Q}$ for any prime number p.

Theorem 11

For any $a > 1$ there exists a positive number x, such that $x^2 = a$.

Sequences

Although the rational numbers are not sufficient to exhaust the points on the number line, sets of rational numbers can be used to exhaust all points in a special way. To be precise, we need to make use of the concept of a *sequence*.

A *sequence of real numbers* is a function from the set of counting numbers (denoted by $\mathbb{Z}^+$) to the real numbers. Suppose that f is such a function; the graph of f consists of pairs $(n, f(n))$, where $n \in \mathbb{Z}^+$. By convention, we consider only the second entries, and write a_n instead of $f(n)$. The sequence is then written as $\{a_n\}$, or $\{a_n\}_{n=1}^{\infty}$.

Definition: A sequence $\{a_n\}$ is said to have limit c if for each positive number ε, there is a positive integer n, such that $|a_k - c| < \varepsilon$ for all $k \geq n$.

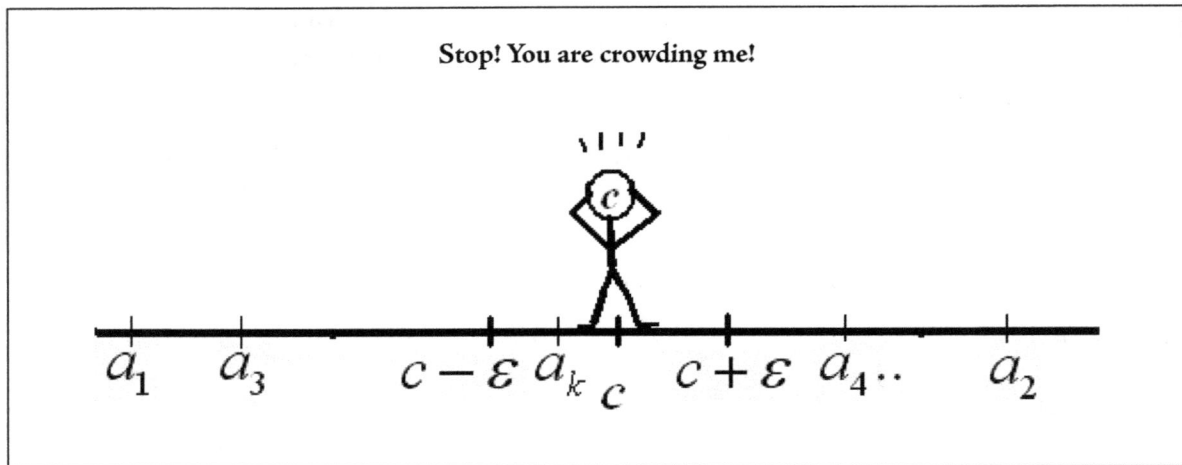

Stop! You are crowding me!

Concerning notation, the statement "$\{a_n\}$ has limit c" is often written as "$\lim_{n\to\infty} a_n = c$" and is often stated as "$\{a_n\}$ converges to c."

Exercise Set 14

Prove each statement and the included theorems:

1. If $\lim_{n\to\infty} a_n = c$ and $\lim_{n\to\infty} a_n = d$, then $c = d$.

2. If $\lim_{n\to\infty} a_n = c$ and $\lim_{n\to\infty} b_n = d$, then $\lim_{n\to\infty} s_n = c + d$, where

$s_n = a_n + b_n$; and $\lim_{n\to\infty} p_n = c \cdot d$, where $p_n = a_n \cdot b_n$.

The next three theorems describe how all real numbers occur on the number line and how they can be approximated by rational numbers.

Theorem 12

If $\{a_n\}$ is a sequence of real numbers, such that $a_k \leq a_{k+1}$ for all $k \in \mathbb{Z}^+$ and for all $n \in \mathbb{Z}^+$, $a_n \leq M$ for some $M \in \mathbb{R}$, then $\lim_{n\to\infty} a_n = c$, where $c = lub\{x \mid x = a_n \text{ for all } n\}$.

Theorem 13 (DENSITY OF $\mathbb{Q}$ IN $\mathbb{R}$)

Suppose that $x, y \in \mathbb{R}$ with $x < y$, then there exists $q \in \mathbb{Q}$, such that $x < q < y$.

Theorem 14

Let r be a real number; then there exists a sequence of rational numbers converging to r.

Definition: A *decimal representation* for a nonnegative real number is an expression of the form $N.a_1a_2a_3\ldots$, where N is a nonnegative integer and $a_i \in \{0, 1, 2, 3, 4, 5, 6, 7, 8, 9\}$.

Exercise Set 15

1. Suppose that $r \geq 0$; show that there exists a bounded nondecreasing set of rational numbers of the form

$$s_n = N + \frac{a_1}{10} + \frac{a_2}{10^2} + \ldots + \frac{a_n}{10^n}$$

   such that s_n converges to the real number r.

2. Find a sequence of rational numbers that converges to $\sqrt{2}$.

Analysis of the Real Line 2—Open Covers

We now present some essential information relating to sets, intervals, and covers, with a goal of establishing a special property regarding the structure of the real numbers.

Let a, b be real numbers, with $a \leq b$.

1. The set of real numbers $\{x \mid a < x < b\}$ is called an *open interval*, and is denoted by (a, b).

2. The set of real numbers $\{x \mid a \leq x \leq b\}$ is called a *closed interval*, and is denoted by $[a, b]$.

   Definition: Let K be a set of real numbers. An *open cover* for the set K is a collection $\mathcal{C}$ of open intervals, such that every point x in K belongs to some open interval in the collection $\mathcal{C}$.

Observe that open covers are fairly general objects. For example,

1. Suppose that $K = \{\pi\}$. An open cover for K is the set consisting of the single interval $\mathcal{C} = \{(1, 5)\}$. Another open cover for K is the set consisting of $\mathcal{C} = \{(1, 4 + \frac{1}{n}) \mid n = 1, 2, 3, 4, \ldots\}$.

2. Suppose that $K = (-1, 1)$. An open cover of K is the set $\mathcal{C} = \{(-2, 0), (-1/2, \sqrt{2})\}$. An open cover for K that uses an infinite number of intervals can also be given (do this as an exercise).

3. If K is the set of integers, the set $\mathcal{C} = \{(k - \frac{1}{3}, k + \frac{1}{3}) \mid k \in \mathbb{Z}\}$ is an open cover for K.

4. Any open interval (a, b) or closed interval $[a, b]$ can be covered by a finite set of open intervals; in fact, a single interval will suffice in either case.

5. Suppose that $K = (0, 1)$. An open cover for K is the set $\mathcal{C} = \left\{ \left(-e, \dfrac{k}{k+1} \right) \middle| k = 1,2,3,... \right\}$.

   Is it possible to use only a finite number of intervals from $\mathcal{C}$ to cover K? Explain.

EXERCISE SET 16

Describe an infinite collection of open intervals, such that two things occur:

i. Each rational number in the interval $[0, 1]$ belongs to at least one open interval in your collection; that is, your collection of open intervals covers the rational numbers belonging to $[0,1]$.

ii. No finite number of intervals in your collection will be enough to cover the rational numbers belonging to $[0, 1]$; that is, if you select any finite number of open intervals from your collection, there will always be at least one rational number that is not inside any of the finite number of open intervals you have selected.

EXPLORATION 3—OPEN INTERVALS

Prove this statement:

> Suppose that $a, b \in \mathbb{R}$ with $a < b$. If a collection $\mathcal{C}$ of open intervals covers $[a,b]$, then some finite subset of open intervals from $\mathcal{C}$ covers $[a, b]$.

> Hint:
a. Let $K = \{x \mid x \in [a, b]$ and a finite number of intervals from $\mathcal{C}$ covers $[a, x]\}$. The set K is not empty; for example, $a \in K$. In fact, **there exists numbers $x \in K$ greater than a**, and the set K has a least upper bound m and $m \leq b$.

b. To show that $m \in K$, observe that $m \in I$ for some open interval I belonging to $\mathcal{C}$. A property of least upper bounds then implies that $[a, m]$ can be covered by I together with a finite set of intervals from $\mathcal{C}$.

c. To show that $m = b$, assume that $m < b$. Because $m \in K$, an argument similar to that used to verify the boldface statement in (a) yields a contradiction.

At this point, you may be asking, "If we can approximate any real number with sequences of rationals, why can't we just work with rational numbers and forget about the reals?" To answer such a question, we ask you to review exercise set 16 in conjunction with exploration 3.

If you engage in exploration 3, you will find that there is a difference in working on a closed interval of real numbers, say, the interval $[0,1]$, as opposed to using only the rational numbers in closed interval $[0,1]$.

In exercise set 16 you are asked to describe an infinite collection of open intervals, such that two things occur:

i.   Each rational number in the interval [0, 1] belongs to at least one open interval in your collection; that is, your collection of open intervals covers the rational numbers belonging to [0, 1].

ii. No finite number of intervals in your collection will be enough to cover the rational numbers belonging to [0, 1]; that is, if you select any finite number of open intervals from your collection, there will always be at least one rational number that is not inside any of the finite number of open intervals you have selected.

In contrast, the content of exploration 3 is the establishment of a special case of a very fundamental theorem about the geometric structure of the real numbers commonly known as the Heine-Borel theorem. The special case established in exploration 3 is commonly stated as follows:

Let a, b be real numbers with $a < b$; then the closed interval $[a, b]$ is *compact*.

What is being asserted by the use of the term "compact" is that regardless of what infinite collection of open intervals is used to cover the interval $[a, b]$ (not just the rational numbers in the interval), there will always be a finite number of open intervals in the collection that covers the interval $[a, b]$.

The concept of compactness is intuitively difficult to absorb, because we are asked to deal with an infinite collection of (unknown) objects from which some finite collection will emerge, even though we may not be able to identify the finite collection. Thus, the theorem *asserts the existence of something, without telling you how to find it*. Can you identify something similar occurring in the axioms for the real numbers?

We have asserted that compactness is a fundamental property of the field of real numbers. We hope that the next exploration will convince you that real numbers are fully needed to describe the physical world using calculus. It may be worthwhile for the reader/learner to review the definition of a *continuous function* on *a set* D of real numbers.

EXPLORATION 4—FUNCTIONS ON THE RATIONALS

Let K be the set of all rational numbers in the closed interval $[1, 3]$, that is, $K = \mathbb{Q} \cap [1, 3]$

1. Describe a function $f : K \to \mathbb{R}$ which is continuous on K and attains neither a maximum value nor a minimum value on K.

2. Describe a function $f : K \to \mathbb{R}$ which is continuous on K, such that $f(1) = 0$ and $f(3) = 2$ and the range is contained in the closed interval $[0, 2]$ but $f(x) \neq 1$ for any $x \in K$.

3. Describe a function $f : K \to \mathbb{R}$ which is differentiable on K, such that $f(1) = 0$ and $f(3) = 0$ but $f'(x) \neq 0$ for any $x \in K$.

An Important Number

In the previous section of this chapter, the Archimedean property was developed, and Archimedes' connection to the number π was noted. One might make the following statement at this point: "Here we have yet another reference to the number π in a mathematics book. Why is π so important?" There is a wealth of literature available that expounds upon the large body of work devoted to the exploration of the number π. The value of π was first investigated as an entity of interest when it was noted by ancient geometers that the ratio of the circumference of a circle to that of the diameter is a constant that had a value of "a little more than three." As mentioned earlier in the section, Archimedes used a unique approximation method in an effort to find an exact value for this ratio, which became known as the number π. This method relied on a continued process of successively inscribing and circumscribing regular polygons about a circle. Archimedes considered the circumference of a circle to be the limit of the perimeter of the inscribed or circumscribed polygons as the number of sides of the polygon is increased without bound. This way of thinking set the groundwork for future methods used by mathematicians to develop calculus. Using this process, Archimedes was able to deduce the familiar value of 22/7 still often used to approximate the value of π today.

Since the time of Archimedes, finding better and better ways to approximate a value for π became a "holy grail" for mathematicians and inspired many individuals to become mathematicians. In 1768, Johann Lambert proved that π was an irrational number, thus showing that an exact value for π can never be found. This result, however, has not diminished the interest in deriving more refined approximations for π that involve finding a value of π to more and more decimal place accuracy. In this regard, π is used currently to test the computing power of modern computers.

Ferdinand von Lindemann, in 1882, was able to show that π falls into the category of numbers called *transcendental numbers*. That is, π is not the root of any polynomial equation with rational coefficients. It is of interest to note that another famous number in mathematics, e, the natural number, shares this distinction (as was proved by Charles Hermite in 1873). Lindemann's work was used to answer the question about constructibility of numbers such as π or $\sqrt{\pi}$. Constructible numbers and the concept of *constructibility* will be explored in chapter four.

3. Beyond the Number Line: Complex Numbers

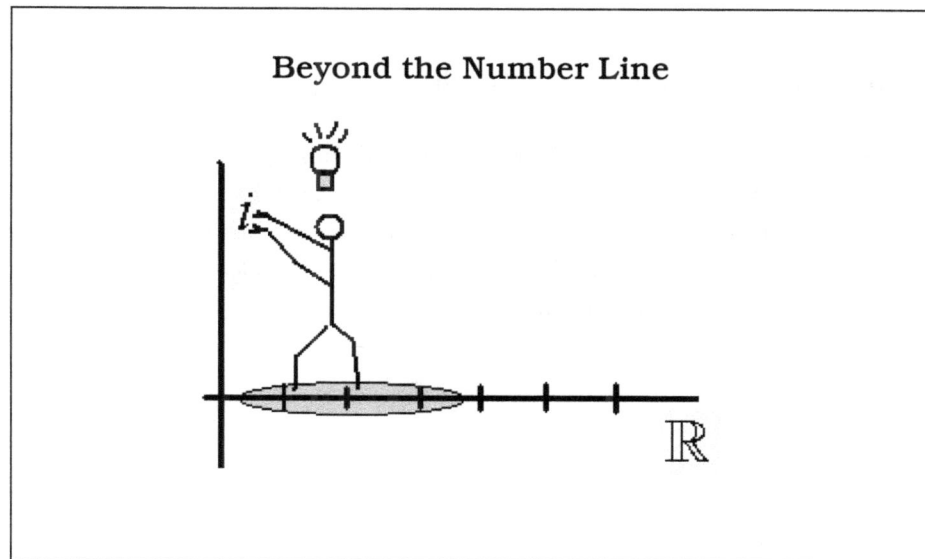

While it is assumed that the learner/reader has had some previous experience with complex numbers, in this chapter we will renew your acquaintance with the geometric properties of complex numbers. A firm understanding of these properties will serve to enhance the axiomatic presentation of the properties of complex numbers offered in chapter four.

The Geometry of Complex Numbers

The *complex number system* is the set generated by the real number system and the imaginary number i. The symbol i is introduced to serve as the solution to the equation

$$x^2 + 1 = 0,$$

and thus is not a real number.

If we consider i as part of a number system, it follows that i^2 is equal to -1. *The complex number z can be defined using ordered pairs* (x, y), *where x and y are real numbers.* The binary operations of addition and multiplication associated with these ordered pairs will be discussed in chapter four. The real numbers x and y are defined as the *real* and *imaginary* parts of the complex number z:

$$z = (x, y).$$

The complex number z can also be written in the form

$$z = x + yi,$$

where x and y are real numbers and i is the imaginary number.

The complex number $z = x + yi$ can be associated with a point in the plane whose coordinates are x and y. We are now moving beyond naming points on the real line and moving into two-dimensional space, or R^2, where points are located in the plane. The x axis in this case is called the *real axis*, and the y axis is named the *imaginary axis*. This is known as the *complex plane*. For example, the number $z = 1 + 3i$ can be represented by the point $(1,3)$ in the complex plane as shown in figure 1.

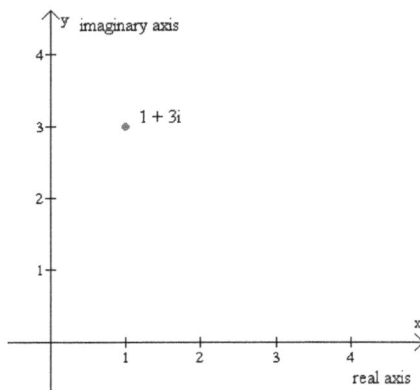

Figure 1

The Geometry of Complex Number Addition

The complex number $z = x + yi$ can also be imagined as a vector from the origin to the point (x, y) as is depicted in figure 2, again, using the number $z = 1 + 3i$.

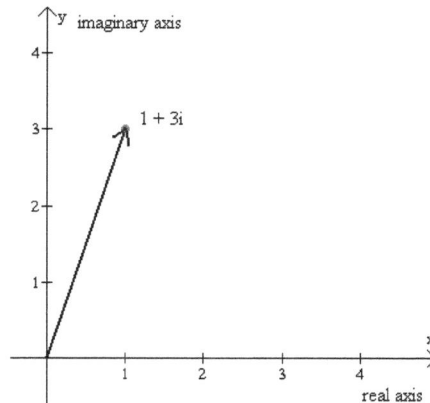

Figure 2

A vector addition approach can be used to find the sum of two complex numbers, $w = x_1 + y_1 i$ and $z = x_2 + y_2 i$. By definition, the sum of w and z is the complex number corresponding to the point $(x_2 + x_2, y_1 + y_2)$, that is, $z + w = (x_1 + x_2) + (y_1 + y_2)i$. Geometrically, the sum of w and z can be found by normal vector addition, as shown in figure 3.

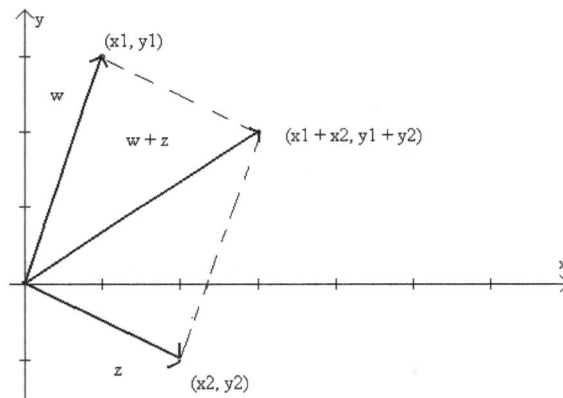

Figure 3

The Modulus of Complex Numbers

The *modulus*, or absolute value of $z = x + yi$, is represented by the symbol $|z|$ and is defined as

$$|z| = \sqrt{x^2 + y^2}.$$

The geometric interpretation of the modulus of z as defined above is that $|z|$ is the distance from the origin to the point (x, y). This distance can be represented simply by a line segment from the origin to the point (x, y), as seen in figure 4.

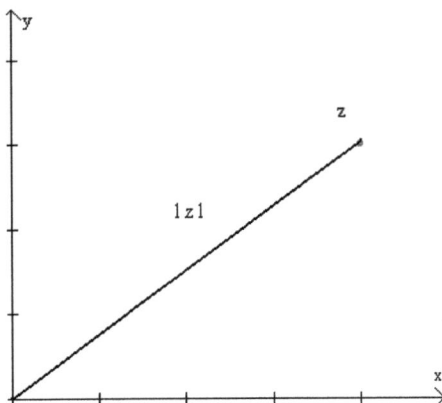

Figure 4

Exercise Set 17

1. If $w = 1 - 4i$ and $z = 2 + 3i$, which point is closer to the origin?

2. Describe the set of all complex numbers that are at a distance of 2 units from the origin.

3. Find the distance between $w = 1 + 2i$ and $z = 3 - i$.

4. Sketch a graph of the set of all complex numbers that satisfy the equation $|z - 2 - 2i| = 3$.

5. Explain the meaning of $|u| < |v|$, where u and v are complex numbers.

The Geometry of Complex Number Multiplication

A complex number $z = x + yi$ can also be written in *polar form*

$$z = r(\cos\theta + i\sin\theta), \tag{1}$$

where r is equal to $|z|$ and θ is defined as the *argument* of z because

$$x = r\cos\theta \text{ and } y = r\sin\theta. \tag{2}$$

The argument θ is found by using the fact that

$$\tan\theta = \frac{y}{x}, \tag{3}$$

where the quadrant containing z must be considered when finding θ. Also note that $\cos\theta + i\sin\theta$ can be written as $e^{i\theta}$ by using *Euler's formula*

$$e^{i\theta} = \cos\theta + i\sin\theta. \tag{4}$$

Therefore, we have

$$z = x + yi = re^{i\theta}, \qquad (5)$$

where the latter form in equation 5 is called the *exponential form*.

The product of two complex numbers z and w is also a complex number in the plane that can be represented by a vector. Unlike the vector description of the geometry of complex number addition, the product zw is not analogous to vector and scalar products of real numbers. This product, however, can be visualized geometrically by considering the multiplication of z and w as written in polar form, as displayed in figure 5.

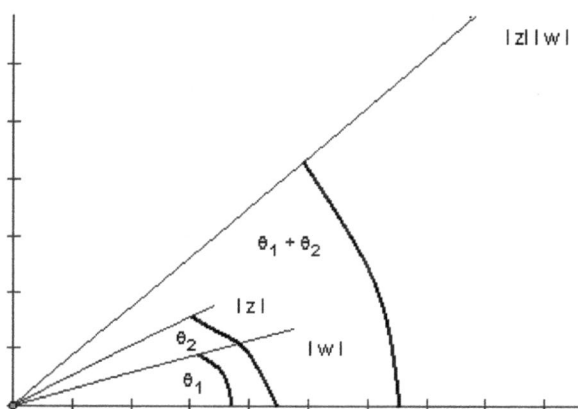

Figure 5

In this chapter we have acquainted (or reacquainted) the learner/reader with the geometric representation and some of the properties of complex numbers. In the next chapter a more formal and axiomatic development of the complex number system will be presented.

Exercise Set 18

Given the complex numbers $z = r_1(\cos\theta_1 + i\sin\theta_1)$ and $w = r_2(\cos\theta_2 + i\sin\theta_2)$,

1. Derive a rule for finding the product zw in both polar and exponential form.

2. Use the result of problem 1 to provide a geometric interpretation of the multiplication of z by i.

3. Derive a rule for finding z^n in both polar and exponential form, where n is an element of the positive integers.

4. Consider the complex numbers u and v. Is it possible to assign meaning to the inequality $u < v$ that extends the order on $\mathbb{R}$?

EXPLORATION 5: NTH ROOTS OF A COMPLEX NUMBER

The result of problem 3 of exercise set 18 is known as De Moivre's theorem. Abraham De Moivre (1667–1754) was a French mathematician who eventually moved to England, where he occasionally collaborated with both Isaac Newton and Edmund Halley. De Moivre's theorem may be used to find the nth roots of a complex number. These n-distinct nth roots are given by

$$r^{\frac{1}{n}}\left[\cos\left(\frac{\theta+2k\pi}{n}\right)+i\sin\left(\frac{\theta+2k\pi}{n}\right)\right],\qquad(\dagger)$$

for $k=\{0,1,2,...,n-1\}$.

1. Use equation † to find the cube roots of $8i$. In addition, plot these roots in the complex plane.

2. Use equation † to find the five fifth roots of unity. In addition, plot these roots in the complex plane.

3. Use equation † to find the solutions of $x^3-7=0$.

4. Use equation † to factor the polynomial $P(x)=x^5-3$ into linear factors. Complex coefficients are permitted.

4. Constructible Numbers: An Interesting Field

Recall that in chapter three we moved off of the real number line and explored numbers that had to be represented as points in the plane. Complex numbers were introduced geometrically as points in the Cartesian plane. In this chapter, we will axiomatically formalize our look at this field of numbers and then explore *constructibility* in relation to numbers in this field.

The Field of Complex Numbers

a.   A *complex number* is an ordered pair (a, b) of real numbers a, b.

b.   The set of all ordered pairs (a, b) where $a, b \in \mathbb{R}$ is denoted by $\mathbb{C}$. We note that two ordered pairs (a, b) and (c, d) are equal if and only if $a = c$ and $b = d$.

c.   Addition and multiplication in $\mathbb{C}$ are defined in the following manner:

$$(a, b) + (c, d) = (a + c, b + d)$$
$$(a, b) \cdot (c, d) = (ac - bd, ad + bc)$$

d.   For any real number r and any $(a, b) \in \mathbb{C}$, let $r \cdot (a, b) = (ra, rb)$. (This is treating the members of $\mathbb{C}$ as vectors, as discussed in chapter three, and the members of $\mathbb{R}$ as scalars.)

e.   Recall that a field is a set K with two operations, usually denoted by $+, \cdot$ (*addition* and *multiplication*), satisfying axioms A1 through A5, M1 through M5, and D.

Theorem 15

With addition and multiplication as indicated in statement c, $\mathbb{C}$ is a field.

Exercise Set 19

1. Show that each of the following is a *field*

    (i) $\mathbb{Q}(i) = \left\{ z \in \mathbb{C} \,\middle|\, z = a + bi, \text{where } a, b \in \mathbb{Q} \right\}.$

    (ii) $\left\{ x \in \mathbb{R} \,\middle|\, x = a + b\sqrt{2} \text{ where } a, b \in \mathbb{Q} \right\}$ with ordinary addition and multiplication (this field is usually denoted by $\mathbb{Q}\left(\sqrt{2}\right)$.

2. In $\mathbb{C}$ let 1 denote the pair $(1, 0)$, and let i denote the pair $(0, 1)$. Verify that:
    $$1 \cdot (a, b) = (a, b) \in \mathbb{C}; -1 = (-1, 0); i^2 = -1; i^3 = -i \text{ and } i^4 = 1$$

3. If $(a, b) \in \mathbb{C}$ show that
    $$(a, b) = a \cdot 1 + b \cdot i \text{ and } (a, b) \cdot (c, d) = (ad - bc)1 + (ad + bc)i$$

Constructible Numbers

At this point we will explore the historically significant concept of *constructibility*. *Constructibility* refers to the determination of those numbers that can be constructed in the plane using only certain rules associated with a compass and straightedge, starting with a *unit segment* that represents the number 1. The goal of this process is to thoroughly devise an algebraic description of those numbers that we can construct. These numbers are known as *constructible numbers*.

The following are axioms for determining points in the plane (i.e., complex numbers) that are said to be *constructible* using a straightedge and compass.

C1: $(0, 0)$ and $(1, 0)$ are constructible.

C2: If (a, b) and (c, d) are constructible, then the line (line segment) connecting (a, b) and (c, d) is constructible.

C3: If (a, b) is constructible and s is a constructible line segment, then the circle with center at (a, b) and radius s is constructible.

C4: Any point obtained in one of the following three ways is constructible: the intersection of two constructible lines; the intersection of a constructible line and a constructible circle; the intersection of two constructible circles.

The distance between any two "marks" or intersections that we are able to achieve using C4 in any manner of repetition is called a *constructible distance*.

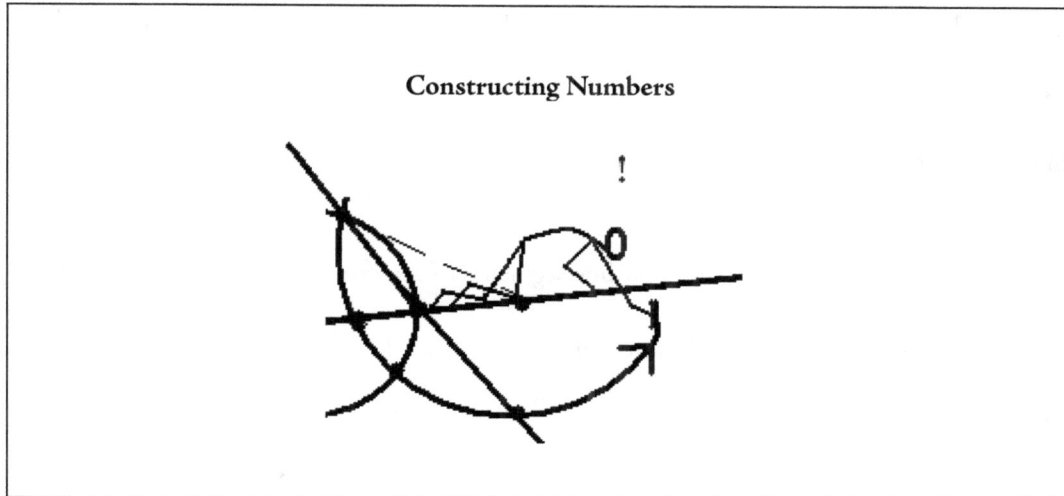

Constructing Numbers

Definition: A complex number $a + bi$ is said to be *constructible* if the point (a, b) is constructible.

Theorem 16

If $\alpha \in \mathbb{C}$ is constructible, then $|\alpha|$ is a constructible distance.

Theorem 17

A complex number $\alpha = a + bi$ is constructible if and only if a and b are constructible real numbers.

Three Famous Problems

Three famous problems originating in ancient Greece involving constructions that mathematicians were not able to accomplish using only the axioms from C1 through C3 are as follows:

1. *The duplication of the cube*, or the problem of constructing the edge of a cube having twice the volume of a given cube.

2. *The trisection of an angle*, or the problem of dividing a given arbitrary angle into three equal parts.

3. *The quadrature of the circle*, or the problem of constructing a square having an area equal to that of a given circle.

It was not until the nineteenth century that the answers to all three of these constructions were obtained. Surprisingly, the solutions to these problems relied not on geometric methods, but on algebra and the use of the mathematics to be outlined in the next chapter.

Exercise Set 20

Prove these statements:

1. The line parallel to a constructible line and passing through a constructible point (not on the line) is constructible.

2. The perpendicular bisector of a constructible line segment is a constructible line.

3. The circle determined by three non-collinear points is constructible.

4. $\sqrt{2} = \left(\sqrt{2}, 0\right)$ is a constructible point.

Theorem 18

If F is the set of all constructible complex numbers, then F is a field.

5.

Constructible Numbers: A Refined Investigation

Are All Real Numbers Constructible?

How big is F, the field of constructible numbers? Does it consist of all complex numbers? Per theorem 17, all complex numbers are constructible if and only if all real numbers are constructible. To begin this chapter, we will concentrate on the constructibility of real numbers.

We will now attempt to establish the connection between constructibility as presented in the previous chapter, the algebraic condition for a real number to be constructible, and the need for introducing the concept of *field extensions*.

Suppose that we start with K, a field of constructible numbers, such that $\mathbb{Q} \subseteq K \subseteq \mathbb{R}$. We wish now to define further points that can be reached using only the compass and straightedge criteria described earlier. Recall that further points can only be found by (a) intersecting two lines, whose coordinates come from K; (b) intersecting a circle and a line, whose coordinates come from K; or (c) intersecting two circles, each of whose coordinates come from K, as depicted in figure 6.

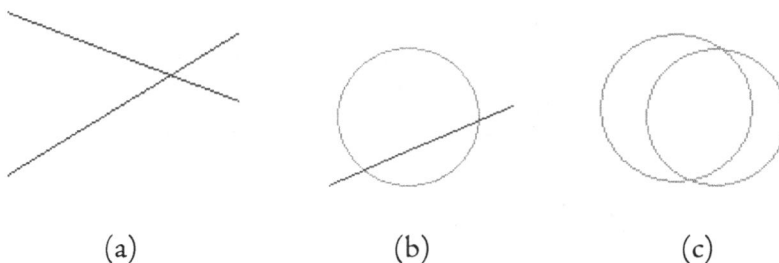

(a) (b) (c)

Figure 6

First, notice that finding a solution to two intersecting lines involves solving a system of two linear equations with coefficients in K; thus, this endeavor will adhere to the field operations from K. It follows that the point of intersection will, in this situation, always have coordinates in K.

Next we consider the intersection of a line and a circle in the plane of K. Let the equation of the line be

$$Ax + By + C = 0, \tag{1}$$

and suppose that the equation of the circle can be written as

$$x^2 + y^2 + Dx + Ey + F = 0, \tag{2}$$

with all coefficients from equations 1 and 2 belonging to K. One could solve for y in equation 1 and substitute this equation into equation 2, yielding

$$x^2 + \left(-\frac{A}{B}x - \frac{C}{B}\right)^2 + Dx + E\left(-\frac{A}{B}x - \frac{C}{B}\right) + F = 0$$

(we assume $B \neq 0$). Notice that the solutions to this quadratic equation will have the form $a + b\sqrt{\alpha}$, where a, b, and α belong to K. Thus, the solutions belong to a quadratic *field extension* of K, namely, $K\left(\sqrt{\alpha}\right)$, and can be constructed.

The fields $\mathbb{Q} \subset \mathbb{Q}\left(\sqrt{2}\right)$, and $\mathbb{Q}\left(\sqrt{2}\right) \subset \left(\mathbb{Q}\left(\sqrt{2}\right)\right)\left(\sqrt{3}\right)$ are examples of field extensions. Thus, for example, we say that $\mathbb{Q}\left(\sqrt{2}\right)$ is field extension of $\mathbb{Q}$.

Last, notice that if we consider the solutions for the intersections of two circles in the plane, we must solve simultaneously two equations having the form of equation 2. Subtracting one of these equations from the other yields a linear equation (which happens to be the equation of the chord in common to the two circles). One could then solve this linear equation simultaneously with either of the original equations of the circles to obtain the coordinates for the points of intersection of the two circles. However, at this point, we note that we are back to dealing with the previous situation of the intersection of a line and a circle. Thus, the solutions will once again be of the form $a + b\sqrt{\alpha}$ belonging to the extension $K\left(\sqrt{\alpha}\right)$.

Exercise Set 21

In exercise set 20, you were asked to show that $\sqrt{2}$ is a constructible number. Now consider the circle with center at the origin and having a radius of $\sqrt{2}$. We can obtain constructible points lying on this circle by intersecting the circle with another constructible circle or with a constructible line.

1. Using only points whose coordinates are in the field $\mathbb{Q}$ of rational numbers, is it possible to construct the circle?

2. Recall (from exercise set 19) that the set $\mathbb{Q}(\sqrt{2})=\left\{x\in\mathbb{R}\,\middle|\,x=a+b\sqrt{2},\text{ where } a,b\in\mathbb{Q}\right\}$ is a field. Is it possible to construct the circle using only points whose coordinates are in this field?

3. Show algebraically that $\left(\dfrac{\sqrt{2}}{2},\dfrac{\left(\sqrt{2}\sqrt{3}\right)}{2}\right)$ are the coordinates of the constructible number obtained

    by intersecting the line $x=\dfrac{\sqrt{2}}{2}$ with the given circle. Based on this result, what can be concluded about $\sqrt{3}$? Is it possible to construct the point of intersection using only points whose coordinates are in this field?

4. Let $K=\mathbb{Q}\left(\sqrt{2}\right)$. Show that $\sqrt{3}\notin K$.

5. Show that $L=K\left(\sqrt{3}\right)=\left\{x\in\mathbb{R}\,\middle|\,x=a+b\sqrt{3},\text{ where } a,b\in K\right\}$ is a field containing $\mathbb{Q}$, $\sqrt{2}$, and $\sqrt{3}$. Is it possible to construct the point of intersection using only points whose coordinates are in this field?

Theorem 19

If $\alpha>0$ is a constructible number, then $\sqrt{\alpha}$ is a constructible number.

Field Extensions and Vector Spaces

We noted earlier that Greek mathematicians posed three geometric construction challenges that they were unable to solve. As it turned out, the mathematics that they had developed up to that time was not sufficient to resolve the problems. What was needed were mathematical methods that went beyond pure geometric constructions. The introduction of algebraic techniques and the reformulation of the problems in algebraic terms provided just such a method.

In this chapter, we will examine this approach to constructible numbers. We have already seen (from exercise set 21) how the geometric process of obtaining constructible numbers such as $\sqrt{2}$ and $\sqrt{3}$ can be reformulated in terms of finding simultaneous solutions of quadratic or linear equations with other quadratic or linear equations. Furthermore, the quadratic equations involve only those arising from equations that represent circles in the plane. In turn, these solutions can lead to the necessity of expanding the field $\mathbb{Q}$ of rational numbers to larger fields in which the solutions can occur, without dealing with the entire field of real numbers.

For example, while we cannot solve the quadratic equation $x^2-2=0$ in the field $\mathbb{Q}$, we can certainly solve the equation in the larger field $\mathbb{Q}\left(\sqrt{2}\right)$, or in any extension field of $\mathbb{Q}$ that contains $\sqrt{2}$. From the point of view of constructible numbers, the field $\mathbb{Q}\left(\sqrt{2}\right)$ is the "best" in which to solve this particular equation, because points having coordinates in $\mathbb{Q}\left(\sqrt{2}\right)$ can serve as starting points for determining other constructible numbers.

The key to determining when a real number is constructible lies in an analysis of how we measure the difference between a field such as $\mathbb{Q}$ and an extension field such as $\mathbb{Q}\left(\sqrt{2}\right)$ or $\mathbb{Q}\left(\sqrt{2}\right)\left(\sqrt{3}\right)$. Such a measure is provided by the concept of *a vector space V over a field K*, and what is called *the dimension of V over K*. Let us proceed to the discussion of these topics.

Formally, a vector space S over a field K is a system involving a set S and a field K for which certain axioms are satisfied. These axioms are as follows:

V1: The set S satisfies axioms A1 through A4.

V2: There is an operation $\cdot$ defined between the elements of K and the elements of S, such that

    i. If $a \in K$ and $x \in S$, then $a \cdot x \in S$

    ii. If $a, b \in K$ and $x, y \in S$, then
$$a \cdot (x + y) = (a \cdot x) + (a \cdot y),$$
$$(a + b) \cdot x = (a \cdot x) + (b \cdot x),$$
$$(a \cdot b) \cdot x = a \cdot (b \cdot x),$$
$$1 \cdot x = x.$$

The members of S are referred to as *vectors*, while the members of K are called *scalars*. The operation of $+$ in S is called *vector addition*, while the operation $\cdot$ between members of K (the scalars) and those of S (the vectors) is called *scalar multiplication*.

Exercise Set 22

Verify:

1. $\mathbb{C}$ is a vector space over $\mathbb{R}$.

2. $\mathbb{R}$ is a vector space over $\mathbb{Q}$.

3. If K, L are fields with $K \subseteq L$, then L is a vector space over K under the operations present in the field.

4. If K is a field, then $K \times K = \left\{(x, y) \mid x, y \in K\right\}$ under the operations
    $(x, y) + (z, w) = (x + z,\ y + w)$, and $a(x, y) = (ax,\ ay)$ is a vector space over K.

5. Generalize the construction in problem 4 to more than two terms.

Linear Independence, Dimension, and Bases

We recall the basic properties of vector spaces involving dimension.

Assume that S is a vector space over a field K, and let $v_1, v_2, \ldots, v_n$ be a finite number of vectors in S.

a. $v_1, v_2, \ldots, v_n$ are *linearly independent* if and only if the only linear combination of $a_1 v_1 + \ldots + a_n v_n$ (where $a_i \in K$) that yields the zero vector is the combination with $a_1 = a_2 \ldots = a_n = 0$ (i.e., $a_1 v_1 + \ldots + a_n v_n = 0 \Rightarrow a_1 = a_2 \ldots = a_n = 0$).

b. The vectors $v_1, v_2, \ldots, v_n$ *span* S if and only if every vector in S can be expressed as a linear combination of the vectors $v_1, v_2, \ldots, v_n$ (i.e., given $v \in S$, there exist scalars $a_1, a_2 \ldots, a_n \in K$, such that $v = a_1 v_1 + \ldots + a_n v_n$).

c. $\{v_1, v_2, \ldots, v_n\}$ is a *basis* for S over K if and only if the vectors $v_1, v_2, \ldots, v_n$ are linearly independent and span S.

d. If S has a basis over K consisting of n vectors, then
    i.   Any basis for S over K has exactly n vectors.
    ii. A set of k vectors where $k > n$ is not a linearly independent set of vectors.

> **Definition:** S has dimension n over K, written $\dim_K S = n$ if it has a basis with exactly n vectors. In this case we say that S is a finite-dimensional vector space over K.

Exploration 6: Independence properties

Prove the previous statements given in d (i) and d (ii).

Now assume that K and L are fields with $K \subseteq L$. As we saw in exercise set 22, we can consider L to be a vector space over K. We say that L is a *finite extension* of K if L is a finite dimensional vector space over K. Thus, L is a finite extension of K if and only if there exists a finite number of elements $\alpha_1, \ldots, \alpha_n$ in L that are linearly independent over K and span L.

> **Definition (Quadratic Extension):** A finite extension L of a field K is called a *quadratic extension* if $\dim_K L = 2$.

Exercise Set 23

Now let $K = L\left(\sqrt{3}\right)$ as in exercise set 21, problem 5.

1. Show that $\{1, \sqrt{3}\}$ are a basis for K considered as a vector space over L. (Hence, K has dimension 2 over L, and K is a quadratic extension of L).

Prove the following:

2. **THEOREM 20:** Suppose that K, L, and F are fields with $K \subseteq L \subseteq F$, such that $\dim_K L = m$ and $\dim_L F = 2$. Show that if $\{x_1, x_2, \ldots, x_m\}$ is a basis for L over K and $\{u, v\}$ is a basis for F over L, then $\{x_1 u, x_2 u, \ldots, x_m u, x_1 v, x_2 v, \ldots, x_m v\}$ is a basis for F over K; hence, $\dim_K F = 2m$.

Theorem 21

If $F \subseteq E \subseteq K$, then $\dim_F K = \dim_F E \cdot \dim_E K$.

Exercise Set 24

If $\mathbb{Q} \subseteq L_1 \subseteq L_2 \subseteq \ldots \subseteq L_m$ is a series of quadratic extensions, determine $\dim_{\mathbb{Q}} L_m$.

Connection to Polynomials

As a review exercise, we will recall the following statements and theorems relating to polynomials.

Exercise Set 25

Prove the following:

Theorem 22 (remainder theorem)

1. Suppose that K is a field. If $p(x)$ and $f(x)$ are polynomials with coefficients in K, then there exist polynomials $q(x)$ and $r(x)$ (with coefficients in K), such that

    (i) $p(x) = f(x) \cdot q(x) + r(x)$
    (ii) Either $r(x) = 0$ or degree $(r(x)) <$ degree $(f(x))$

    (Note: A nonzero polynomial $g(x)$ has degree t if x^t is the highest power of x occurring in $g(x)$ with a nonzero coefficient.)

Theorem 23 (the rational root theorem)

2. If $f(x) = a_0 + a_1 x + \ldots + a_n x^n$ is a polynomial with integral coefficients and p/q is a rational root where $\gcd(p, q) = 1$, then p divides a_0 and q divides a_n.

The following important theorems are introduced one at a time. You are encouraged to prove each one of them. Note that a polynomial $f(x) \in F[x]$ is irreducible if f cannot be written in the form $f(x) = g(x) \cdot h(x)$, where $g(x)$ and $h(x)$ have a degree less than $f(x)$.

Suppose that L is a finite extension of K, with $\dim_K L$.

Theorem 24

If $\alpha \in L$, then there exists a polynomial

$$p(x) = x^m + a_{m-1}x^{m-1} + \ldots + a_1 x + a_0 \text{, where } a_i \in K \text{, such that } 1 \le m \le n \text{ and}$$
$$p(a) = x^m + a_{m-1}a^{m-1} + \ldots + a_1 a + a_0 = 0.$$

Theorem 25

If $p(x)$ is a polynomial as in theorem 22, and $p(x)$ is chosen so that m is minimal, then m is a divisor of $n = \dim_K L$.

Theorem 26

Suppose that K is a finite extension of F and $\alpha \in K$ is the root of an irreducible polynomial $f(x)$, an element of $F[x]$ of degree n. Then $\dim_F F(\alpha) = n$.

Theorem 27

Suppose $\dim_F K = n$ and $\alpha \in K$ is a root of an irreducible polynomial $f(x) \in F[x]$. Then $deg(f(x)) \mid n$.

And Now...The Big Finale!

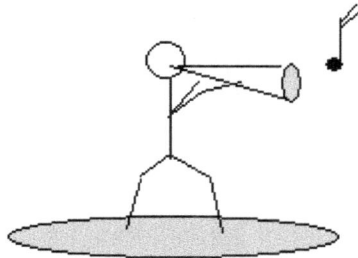

The Big Finale—For Now

Theorem 28

A real number α is constructible if and only if there exists a field K where $\mathbb{Q} \subseteq K \subseteq \mathbb{R}$, such that $\alpha \in K$ and $\dim_{\mathbb{Q}} K = 2^t$ for some integer $t \geq 0$.

Theorem 29

There is no general construction by straightedge and compass that trisects an angle.

Outline of proof:

a. Use the fact that $\cos(a + b) = \cos a \cos b - \sin a \sin b$ to show that

$$\cos 3t = 4\cos^3 t - 3\cos t . \tag{1}$$

b. Let $3t = 60°$ in equation 1. Write the result.

c. Use a substitution in the result of statement b above to create a cubic polynomial. What conclusions can you make concerning α, a root of the cubic?

d. Based on your conclusions about α, what is $\dim_{\mathbb{Q}} \mathbb{Q}(\alpha)$? What does this imply about the constructibility of α.]

(Note: **For further thought,** determine whether all roots of

$$x^4 - 2x^3 - 2x + 4$$

are constructible.)

6.

A Progression of Thought: A Connected Overview

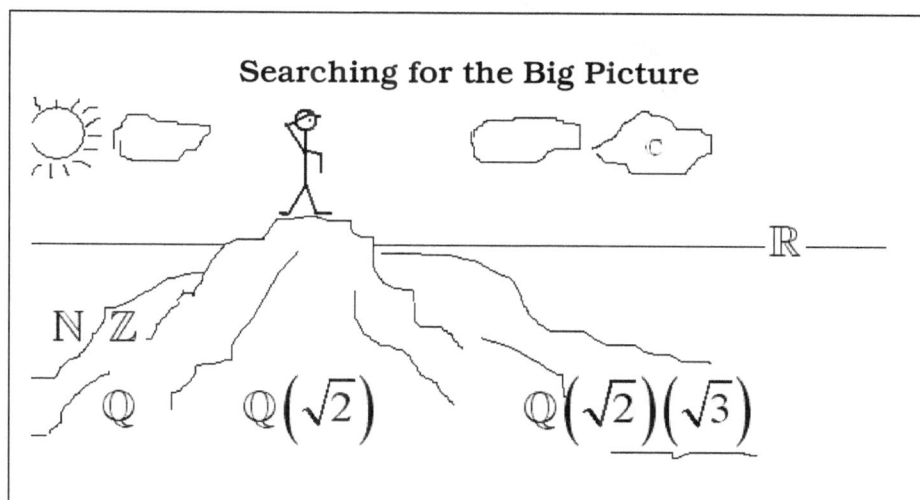

Searching for the Big Picture

The Algebraic Formulation of the Integers, Rational Numbers, Real Numbers, and Complex Numbers (Finite Fields)

The primary function of algebra is finding the solutions of equations that involve "numbers." "Numbers" is a very loose concept—in general, we are only permitted to "add" and "multiply" numbers subject to certain restrictions.

I. The most basic system of "numbers," together with "addition" and "multiplication," is the set of *counting numbers* $\{0, 1, 2, 3, \ldots\} = \mathbb{N}$, with ordinary addition and multiplication. An example of an equation using numbers from $\mathbb{N}$ is the equation

$$x + 5 = 0.$$

This equation has no solution belonging to the set $\mathbb{N}$. In order to solve an equation of the form $x + a = 0$, where $a \in \mathbb{N}$, we expand our number system by introducing the symbol $-a$, for each $a \in \mathbb{N}$:

$$-a \text{ is the solution to } x + a = 0.$$

As a result, we obtain the set $\mathbb{Z}$ of integers: $\mathbb{Z} = \{0, 1, 2, 3, \ldots\} \cup \{-1, -2, -3, -4, \ldots\}$

Of course, we must check that our rules for addition and multiplication inherited from $\mathbb{Z}$ also apply to $\mathbb{Z}$.

II. While we can now solve equations of the form $x + a = 0$, as well as others, such as $2x + 4 = 0$, using only properties of $\mathbb{Z}$, we cannot solve an equation such as $2x - 3 = 0$. That is, there does not exist an integer a such that $2a - 3 = 0$, or, equivalently, such that $2a = 3$.

As a result, we introduce solutions to equations of the form $ax = b$, where $a,b \in \mathbb{Z}$ and $a \neq 0$ and the rational numbers $\mathbb{Q}$ are created. *But*...some difficulties must be overcome (annoyances, really). For example, if $a \neq 0$ is an integer, we define a^{-1} as the number, such that $a^{-1} \times a = 1$. Equivalently, a^{-1} is the solution to the equation $ax = 1$, which has no solution in $\mathbb{Z}$ if $|a| < 1$

Another equation with no solution in $\mathbb{Z}$ is the equation

$$6x = 3.$$

Experience tells us that this is equivalent to solving the equation $2x = 1$.

Exercise Set 26

1. Using properties of $\mathbb{Z}$, show that for $a,b,c \in \mathbb{Z}$ with $a \neq 0$, the equation $(ab)x = c$ is solvable in $\mathbb{Z}$ if and only if $bx = c$. (Note: The key property of $\mathbb{Z}$ is that if $uv = 0$, then $u = 0$ or $v = 0$).

Returning to the equation $6x = 3$ is the solution $6^{-1} \times 3$ or $2^{-1} \times 1$. Remember that 6^{-1} and 2^{-1} are symbols having certain properties: $6^{-1} \times 6 = 1$, $2^{-1} \times 2 = 1$. We would expect $6^{-1} \times 3$ to be the same number as $2^{-1} \times 1$, although the representations may look different. We also want to add and multiply expressions such as $6^{-1} \times 3$ and $5^{-1} \times 10$ without having to deal with what representation we use; that is, we want the results of our calculations to be independent of the representations we choose to use for our calculations.

For these reasons, we agree that two expressions $b^{-1}a$ and $d^{-1}c$ are *equivalent* if and only if $bc = ad$. Convention also leads to writing the expression $b^{-1}a$ as a fraction a/b, and the statement that fractions are equivalent now is written as $a/b = c/d$ if and only if $ad = bc$. Once equivalent fractions are defined, we can define addition and multiplication as follows:

$$\frac{a}{b} + \frac{c}{d} = \frac{ad + bc}{bd} \text{ and } \frac{a}{b} \cdot \frac{c}{d} = \frac{ac}{bd}.$$

Exercise Set 27

Verify that operations are *well defined*; that is, if a/b and a_1/b_1 are equivalent, and c/d and c_1/d_1 are equivalent, then $a/b + c/d$ is equivalent to $a_1/b_1 + c_1/d_1$, and so on.

Note that the set $\mathbb{Q}$ of all equivalent fractions is the field of rational numbers, and the integers are represented by the fractions equivalent to $a/1$, where $a \in \mathbb{Z}$.

Having constructed rational numbers from the integers as a means to solve linear equations, we can follow similar paths to construct other fields as a means to solve quadratic equations. We need to do this because, for example,

1. If we have the real numbers, then $\sqrt{2}$ exists.

2. $\sqrt{2}$ is not a rational number.

Put another way, there exists a real number a, such that $a^2 = 2$, but a is not a rational number. In algebraic terms,

$$\text{the equation } x^2 - 2 = 0 \text{ has no solution in } \mathbb{Q}.$$

In terms of polynomials, the polynomial $x^2 - 2$ cannot be written as a product of two linear factors whose coefficients are in $\mathbb{Q}$.

EXERCISE SET 28

Prove the following:

The polynomial $p(x)$ has a linear factor $x - a$ if and only if the polynomial equation $p(x) = 0$ has a as a solution.

Arguing as before we can construct a field $\mathbb{Q}(\sqrt{2})$ in which we can solve the equation $x^2 - 2 = 0$ as follows. Because we want a number θ such that $\theta^2 = 2$, but we also want to keep the rational numbers, it makes sense to introduce a solution θ to the equation $x^2 - 2 = 0$ and consider all expressions of the form

$$a + b\theta, \text{ where } a, b \in \mathbb{Q}.$$

We agree that two expressions

$$a + b\theta \text{ and } c + d\theta$$

are equivalent if and only if $a = c$ and $b = d$.

We then define addition and multiplication using the rules for distribution and addition and multiplication in $\mathbb{Q}$ plus the fact that θ is a solution to the equation $x^2 - 2 = 0$:

$$(a + b\theta) + (c + d\theta) = (a + c) + (b + d)\theta$$

$$(a + b\theta)(c + d\theta) = (ac + 2bd) + (ac + bc)\theta$$

Exercise Set 29

1. An equivalent formulation of the equivalence of expressions is to define how 0 is represented: $A + B\theta = 0$ if and only if $A = 0$ and $B = 0$.

2. Show that addition and multiplication in $\mathbb{Q}\,(\theta)$ are well-defined operations.

3. Show that the polynomial $x^3 - 3x - 1$ has no roots in $\mathbb{Q}$.

4. Use the polynomial from the previous problem to construct a field extension K of $\mathbb{Q}$ such that $\dim_{\mathbb{Q}} K = 3$ and $x^3 - 3x - 1$ has a root in K.

Exercise Set 30

Construct (algebraically) the complex numbers $\mathbb{C}$ from the real numbers $\mathbb{R}$.
(You will need a quadratic polynomial that has no solution in $\mathbb{R}$).

Countability

We will start this section with a definition of what is meant by the expression "an infinite set may be countable."

> **Definition:** Let S be an infinite set; then S is *countable* if there exists a one-to-one function from S into the set $\mathbb{N}$ of natural numbers.

According to the definition, any infinite subset of $\mathbb{N}$ is countable. Thus, infinite countable sets are sets that can be placed in an *injective* correspondence with an infinite subset of $\mathbb{N}$. One consequence of the well-ordering property of the real numbers is that the set $\mathbb{N}$ is itself well ordered: every non-empty subset of N has a least (first) element (see page 61). This property can be used to establish the next theorem, which shows that an infinite subset of $\mathbb{N}$ can be placed in a *bijective* correspondence with $\mathbb{N}$. Thus, in some sense, $\mathbb{N}$ is the "smallest" infinite set.

Theorem 30

If T is an infinite subset of $\mathbb{N}$ then there exists a bijection $g : \mathbb{N} \to T$. Thus, an infinite set S is countable if and only if there is a bijection $h : S \to \mathbb{N}$.

The effect of theorem 30 is that the elements of an infinite countable set can labeled by the natural numbers; that is, the elements can be listed as a first element, second element, third element, and so forth, and all elements of the set occur in the list.

An infinite set that is not countable is said to be *uncountable*. Another way to think of a set S being uncountable is that there cannot exist a one-to-one function from $\mathbb{N}$ to S that is also onto. In other words, any function from S to $\mathbb{N}$ will not be one-to-one.

Exercise set 31 gives an alternate approach to countability.

Exercise Set 31

1. Consider the function $f : \mathbb{Q}^+ \to \mathbb{N}$ defined by:

    for $q > 0$, $q = \dfrac{a}{b}$ where $a, b > 0$ integers and $(a, b) = 1$ let $f(q) = 2^a 3^b$.

    a   Verify that f is a 1-1 function.

    b.   What does this show about the set $\mathbb{Q}^+$?

The basic idea behind countability is that a set S is countable if the elements of S can be "labeled" by the positive integers; that is, S can be described in the form

$$S = \{x_1, x_2, x_3, x_4 \ldots\}.$$ (This is the essence of **Theorem 30**)

2. Suppose $\{A_1, A_2, A_3, A_4 \ldots\}$ is a countable set, where each A_i is a countable set.

    "Double label" each set as

    $$A_i = \{x_{i1}, x_{i2}, x_{i3}, x_{i4} \ldots\},$$

    and let

    $$A = \bigcup_{i=1}^{\infty} A_i \qquad \left(\text{where} \bigcup_{i=1}^{\infty} A_i = \text{disjoint union of the sets } A_i \right).$$

    Can you define a one-to-one function from A to $\mathbb{N}$, as in problem 1 of this exercise set?

Theorem 31

A countable union of countable sets forms a countable set.

Exercise Set 32

1. Suppose that $p_0, p_1, p_2, \ldots, p_k$ are distinct primes, and let $n_0, n_1, n_2, \ldots, n_k$ be integers (not necessarily positive). Show that if $p_0^{n_0} \cdot p_1^{n_1} \cdot p_2^{n_2} \cdots p_k^{n_k} = 1$, then $n_0 = n_1 = \ldots = n_k = 0$.

2. Let S be the set of all polynomials in x with integer coefficients; a typical member of S looks like $a_0 + a_1 x + a_2 x^2 + \ldots + a_n x^n$. Let $P = \{p_0, p_1, p_2, p_3, \ldots\}$ denote the set of all primes. Use a similar idea to that in problem 1 (or 2) of exercise set 31 to define a function f from $S \to \mathbb{Q}^+$ that makes use of problem 1 of this exercise set to verify that f is one-to-one.

   What can you conclude about S? (And why?)

3. (Related to problem 2). A polynomial of degree n (with integer coefficients) has at most n distinct roots. For each $n \geq 1$, show that

$$A_n = \left\{ z \in \mathbb{C} \,\middle|\, \begin{array}{l} z \text{ is a root of an irreducible polynomial with integer} \\ \text{cofficients having degree } n \end{array} \right\}$$

   is a countable set.

We now define what is meant by the term *algebraic number*.

Definition: A number is *algebraic* if it is a root of a polynomial equation with rational coefficients.

Recall that at the end of chapter two, it was noted that Ferdinand von Lindeman was able to show that numbers such as π and $\sqrt{\pi}$ were *transcendental numbers*. A real number is said to be *transcendental* if it is not algebraic.

"Transcendental" Meditation

Theorem 32

The set of algebraic real numbers is a countable set.

Exploration 7

The Real numbers on [0, 1]

Show that the interval [0, 1] is an uncountable set of real numbers.

Theorem 33

$\mathbb{R}$ is an uncountable set.

Exercise Set 33

Show that the field of real constructible numbers is a countable set, and, therefore, not all real numbers are constructible.

We hope that you have convinced yourself at this point that theorems 32 and 33 are true statements. The set of real numbers forms an uncountable set, and the set of algebraic real numbers is a countable set. What does this imply about the countability of the transcendental real numbers?

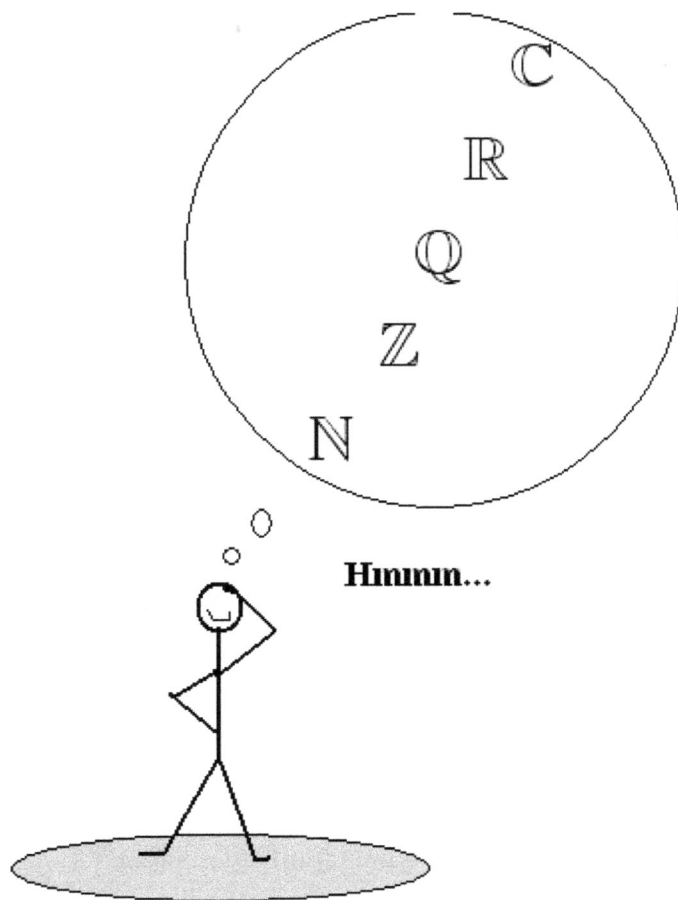

A Progression of Thought

7. Finite Fields

The last topic we will present is the construction of fields having only a finite number of elements. The construction requires elementary concepts from number theory; we summarize these concepts in the next few sections.

Division and the Well Ordering Principle

Let $a, b \in \mathbb{Z}$ with $a \neq 0$. We say that a *divides* b if $b = ka$ for some $k \in \mathbb{Z}$. We write $a|b$ to indicate that a divides b. If $a|b$ then a is called a *divisor* of b.

Exercise Set 34

Determine the validity of each of the following statements, providing appropriate proofs or counterexamples:

Let $a, b, c \in \mathbb{Z}$ with $a \neq 0$

1. If $a|b$ and $b|c$ then $a|c$.

2. If $a|bc$, then $a|b$.

3. If $a|b^2$, then $a|b$.

4. If $a|b$, then $a|bx$ for any integer x.

5. If $a|b$ and $a|c$, then $a|bx + cy$ for any integers x, y.

A fundamental property that $\mathbb{Z}$ possesses is the *well-ordering principle*.

If S is a set of integers which is bounded below (i.e., there exists $M \in \mathbb{Z}$ such that $M \leq x$ for all $x \in S$), then S contains a smallest integer.

The well-ordering principle can be used to prove the following:

Theorem 34

Let $a, b \in \mathbb{Z}$ with a > 0. Then there exists $q, r \in \mathbb{Z}$, such that

$$b = aq + r \text{ and } 0 \le r < a.$$

Let $a, b \in \mathbb{Z}$ not both zero; $d \in \mathbb{Z}$ is the *greatest common divisor* of a and b if

(i) $d|a$ and $d|b$

and   (ii) if $c \in \mathbb{Z}$, such that $c|a$ and $c|b$, then $c \le d$.

The greatest common divisor of two integers a and b is denoted by the symbol (a, b).

Theorem 35

Let $a, b \in \mathbb{Z}$ not both zero, and let $d = (a, b)$. Then there exist $u, v \in \mathbb{Z}$, such that $d = au + bv$. In fact, d is the smallest positive integer that can be expressed in the form of $as + bt$ for some integers s and t.

Prime Numbers

An integer p > 1 is prime if its only positive divisors are 1 and p.

Theorem 36

Let p be any prime integer. Then for any $a \in \mathbb{Z}$, $(a, p) = 1$ if $p \nmid a$, while $(a, p) = p$ if $p | a$.

Theorem 37

Let p be a prime integer, and $a, b \in \mathbb{Z}$. If $p | ab$, then $p | ab$ or $p | b$.

Congruence Modulo *m*

Let m be a positive integer. Given integers $a, b \in \mathbb{Z}$ we say that *a is congruent to b modulo m* if $m|(a-b)$. In symbols

$$a \equiv b \bmod m$$

if and only if $m|(a-b)$.

Exercise Set 35

1. Let $m \in \mathbb{Z}$, $m > 0$, and let $a, b, c \in \mathbb{Z}$. Show that

   (i) $a \equiv a \bmod m$

   (ii) If $a \equiv b \bmod m$ then $b \equiv a \bmod m$

   (iii) If $a \equiv b \bmod m$ and $b \equiv c \bmod m$ then $a \equiv c \bmod m$.

   (iv) If $a \equiv b \bmod m$ then $a + c \equiv b + c \bmod m$ and $ac \equiv bc \bmod m$

2. If p is a prime integer and $a \in \mathbb{Z}$, then $a \equiv r \bmod p$ where $r \in \{0, 1, 2, \ldots, p-1\}$.

3. Suppose that p is a prime integer. If $(a, p) = 1$ then there exists $b \in \mathbb{Z}$ such that $ab \equiv 1 \bmod p$.

Theorem 38 (Fermat's little theorem)

If p is a prime integer and $a \in \mathbb{Z}$ with $(a, p) = 1$, then

$$a^{p-1} \equiv 1 \bmod p.$$

The Finite Field $\mathbb{Z}_p$

Given a prime integer p, let $\mathbb{Z}_p = \{0, 1, 2, \ldots, p - 1\}$. We define addition and multiplication in $\mathbb{Z}_p$ using "congruence modulo p." That is,

   (1) if $a, b \in \mathbb{Z}$ then $a + b = s \in \mathbb{Z}_p$
      where $s \equiv (a + b) \bmod p$

   (2) if $a, b \in \mathbb{Z}$ then $a \cdot b = m \in \mathbb{Z}_p$
      where $m \equiv (a \cdot b) \bmod p$.

Exercise Set 36

Construct addition and multiplication tables for $\mathbb{Z}_5$.

Theorem 39

If p is a prime integer, then $\mathbb{Z}_p$ is a field with respect to addition and multiplication modulo p defined in (1) and (2),.

Exercise Set 37

Let p be a prime integer and assume K is a field having exactly p elements.

1. If $\mathbf{1}$ is the multiplicative identity for K, show that $K = \{n \cdot \mathbf{1} \mid n = 0, 1, \ldots, p-1\}$.

   [Note: If $n > 0$, $n \cdot \mathbf{1} = \mathbf{1} + \ldots + \mathbf{1}$ (n terms), and $0 \cdot \mathbf{1} = 0$ (the additive identity of K).]

2. If ϕ is the function from $\mathbb{Z}_p$ to K defined by $\phi(n) = n \cdot \mathbf{1}$, then ϕ is one-to-one and onto. Furthermore $\phi(nm) = \phi(n) \cdot \phi(m)$ and $\phi(n+m) = \phi(n) + \phi(m)$. ($\varphi$ is then an *isomorphism* from $\mathbb{Z}_p$ onto K; $\mathbb{Z}_p$ and K are said to be *isomorphic* to each other.

3. If $p \geq 5$, show that $\left\{ b^2 \mid b \in \mathbb{Z}_p \right\} \neq \mathbb{Z}_p$.

Theorem 40

Let p be a prime integer and suppose $a \in \mathbb{Z}_p$ such that $a \uparrow b^2$ for all $b \in \mathbb{Z}_p$. Then

(i) The equation $x^2 + a = 0$ has no solution in $\mathbb{Z}_p$.

(ii) There is a field K containing $\mathbb{Z}_p$ such that K has exactly p^2 elements and the equation $x^2 + a = 0$ is solvable in K.

Theorem 41

If K is a finite field, then K contains a subfield H, such that H has exactly p elements for some prime integer p. Furthermore, the number of elements in K is p^n for some integer $n \geq 1$.

Exercise Set 38

1. Construct a field with exactly 25 elements.

2. Construct a field with exactly 125 elements.

Exploration 8: Constructing A Finite field

Outline a procedure for constructing a field having exactly p^n elements, for any prime integer p, and for any integer $n \geq 1$.

8. Probability: From a Discrete to a Continuous Function

In chapter two, we introduced the completeness axiom of the real numbers and the idea of compactness to "fill in" the real number line. The completeness of the real numbers is essential to working with continuous functions that map the real numbers to the real numbers. As was witnessed in exploration 4, if the property of completeness of the real numbers did not exist, many of the foundational theorems of calculus would fail, and the idea of continuity would fall apart when applied to functions generally. The completeness of the real numbers and compactness as applied to intervals of real numbers is one way to develop the idea of continuity of functions in the plane. In essence, we start with discrete values or intervals on the number line and use completeness and compactness to create the continuity of real numbers that are then mapped to another real number line as continuous functions.

EXERCISE SET 39

Prove the following:

Statement A1: If f is a continuous function defined on $[a, b]$, then $\{y \mid y = f(x)$ for some $x \in [a,b]\}$ has an upper bound (and a lower bound).

Outline of proof: Use the $\varepsilon - \delta$ definition of f being continuous at each $x \in [a,b]$ with $\varepsilon = 1$; that is, given $x \in [a,b]$, there is a $\delta_x > 0$, such that $|x - z| < \delta_x$ implies $|f(x) - f(z)| < 1$. The set C of intervals $(x - \delta_x, x + \delta_x)$ covers $[a, b]$. Statement A1 permits the construction of an upper bound (and a lower bound).

Another way to start with discrete values or elements belonging to a set and work toward creating a continuous domain for a continuous function can be achieved using probability. This is accomplished by first defining a general probability density function based on a necessary and sufficient set of axioms. The probability function can then be further developed as a discrete function for a sample space with a finite or countably finite number of outcomes or as a continuous function by considering a probability function over a continuous sample space. This development is the main subject of this chapter.

Some Background on Probability: Counting

In order to discuss the concepts of *sample space* and *event*, one must first understand what is meant by the terms *experiment* and *sample outcome*. A statistical *experiment* can be thought of as any process that produces a set of data called the *sample outcomes*. The set of all possible outcomes of an experiment is known as the *sample space*. Thus, for a given experiment, any individual outcome s is a member of the associated sample space S, written $s \in S$. An *event* is, therefore, defined as a subset of the sample space. This subset can consist of any collection of sample outcomes, including the null set or the entire set.

EXERCISE SET 40

1. Characterize (that is, write in set notation) the sample space S of the set of all ordered pairs (x, y) on the boundary or the interior of a circle of radius 3 centered at the origin.

2. Consider the equation

    $$x^2 + 4bx + c = 0.$$

    Let C be the event "this equation has complex roots." Characterize the event C as a set of ordered pairs (b, c).

3. For the sample space $L = \{t | t \geq 0\}$, where t is the life span of a digital television circuit board component, characterize the event E that the digital component fails to operate before the end of the eighth year.

There is often an element of chance that is associated with the occurrence of certain events that arise when one performs an experiment. In order to deal with this situation, we shall use some fundamental principles that apply to *counting* points in a sample space when it is not feasible or desirable to devise a list of each element belonging to a sample space. The first of these principles is the *multiplication rule*.

Theorem 42 (the multiplication rule)

If an operation can be performed in n_1 ways, and if for each n_1 a second operation can be formed in n_2 ways, and so forth, then the sequence of k operations, where k is an element of the positive integers, can be performed in $n_1 \cdot n_2 \cdot \ldots \cdot n_k$ ways.

We may often be interested in a sample space containing elements that consist of all possible arrangements of groups of objects. An example of this arises when one considers how many ways in which two numbers between 0 and 9, along with four letters, can be arranged in order to generate license plates for use on automobiles. These types of arrangements are known as *permutations*.

Definition: A *permutation* is an arrangement of all or part of a set of elements.

Theorem 43

For n distinct elements, the number of permutations possible is $n!$.

Furthermore, if the number of distinct elements n is chosen r at a time, we can state the following:

Theorem 44

For n distinct elements chosen r at a time, the number of permutations possible is written $_nP_r$, where

$$_nP_r = \frac{n!}{(n-r)!}.$$

It is important to note that the arrangements of elements within an outcome set is important to the consideration of the number of permutations of a set of elements. For example, if we are asked how many ways one could choose a panel consisting of a president, vice president, and a secretary from a collection of people p_i, the order within the sets chosen is important. The set {person$_1$ = president, person$_2$ = vice president, person$_3$ = secretary} is different from {person$_2$ = president, person$_3$ = vice president, person$_1$ = secretary}. In this type of situation, "order" clearly matters.

In many problems involving the probability of outcomes, "order" does not matter. That is, the outcome {1, 2, 3} would not be considered different from the set {2, 1, 3}. The set {2, 1, 3} would be viewed as a repetition. This situation arises, for example, when one is simply trying to match numbers drawn in a lottery. If, in a lottery, three numbers are drawn from the set {1, 2, ..., 20}, and the winning numbers are 3, 5, and 6, you would win if you had a ticket that consisted of the three numbers in the order 3, 6, 5 or if your ticket read 3, 5, 6. Order is not important, so the set {3, 5, 6} is not considered a distinct outcome from, for example, {3, 6, 5}. Often, selections arise where r elements are taken from n distinct objects without regarding order in the outcome events. These selections are called *combinations*.

Theorem 45

For n distinct elements chosen r at a time, the number of combinations possible is written $_nC_r$ or $\binom{n}{r}$, where

$$\binom{n}{r} = \frac{n!}{r!(n-r)!}.$$

Note that the symbol $\binom{n}{r}$ is read as "n choose r." The values that are found by evaluating $\binom{n}{r}$ are called *binomial coefficients*, because these values appear in the expansion of a binomial sum raised to a power, as in

$$(x+y)^n = \sum_{r=0}^{n} \binom{n}{r} x^r y^{n-r}.$$

This statement is known as the *binomial theorem*.

Exercise Set 41

Prove the following:

Theorem 46 (the binomial theorem)

For all positive integers n and x, y belonging to the real numbers,

$$(x+y)^n = \sum_{r=0}^{n} \binom{n}{r} x^r y^{n-r} \cdot \binom{n}{r} h^{n-k} t^k$$

As a data model, the binomial theorem can be applied nicely to situations in statistics where each outcome that arises in a sample space is a sequence of n independent trials, each consisting of one of only two possible outcomes—success or failure. This situation will be further investigated later in this chapter with regard to *probability functions*. However, we note that in any two-outcome experiment with probabilities h and t as outcomes $h + t = 1$ and

$$\binom{n}{k} h^{n-k} t^k$$

is called a *binomial probability* for subsets of k elements chosen from n elements.

Exercise Set 42

1. What coefficient will yield the coefficient of the term $x^{10} y^5$ in the expansion of $(x+y)^{15}$?

2. Find the fifth term of the expansion of $(a+b)^{11}$.

3. Expand $(5+i)^6$ using the binomial theorem.

4. Expand $(3+i)^5$, where i is the imaginary number, using the binomial theorem.

5. In Pascal's triangle, add the numbers across each row and create a sequence of these numbers. Do you notice a pattern in this sequence?

$$\sum_{k=0}^{n}\binom{n}{k}=2^{n}$$

6. Prove that for all integers $n \geq 0$,

$$\sum_{k=0}^{n}\binom{n}{k}=2^{n}.$$

7. Verify the following two identities and describe how they relate to Pascal's triangle. For all $r \leq n$,

a. $\binom{n}{r}=\binom{n}{n-r}$

b. $\binom{n}{r-1}+\binom{n}{r}=\binom{n+1}{r}$

8. Use the concept of binomial probability to solve the following problem: an assembly line that produces Christmas bulbs is found to have problems. If the probability is 0.02 that a particular bulb produced by the assembly line is defective, what is the probability that an 18-bulb carton produced by this assembly line will contain

a. Exactly 2 defective light bulbs?
b. More than 3 defective light bulbs?

Sample Spaces and Events

We now turn our attention to defining the *complement* of an event, along with defining some operations with events that result in the formation of new events. These operations will be essential to our development of a certain type of function called *the probability function*.

> **DEFINITION (the complement of an event):** The compliment of an event A with respect to the sample space S is the subset of all elements of S that are not in A. The compliment of A is written as A^{C}.
> For the next three definitions, let A and B be any two events that are defined over a sample space S.

> **DEFINITION (the union of A and B):** The union of events A and B, denoted by $A \cup B$, is the event whose outcomes are elements of either A or B.

> **DEFINITION (mutually exclusive sets) :** The events A and B over the sample space S are *mutually exclusive* (also known as *disjoint*) if $A \cap B = \emptyset$. This means that events A and B have no elements in common.

Probability and Probability Functions

We will now embark on the task of assigning a *probability* to an experimental outcome or event. We will draw heavily on the topics of unions, intersections, and complements from set theory defined in the previous section.

In an experiment having N different equally likely outcomes, with n of these outcomes corresponding to an event A, then the probability P of event A is

$$P(A) = \frac{n}{N}.$$

More generally, in an experiment we can view the probability that each of the various events occurs in a finite sample space S as a function P that is governed by a set of three axioms or postulates. Further, if S has an infinite number of members, then the behavior of P is fully explained by only four axioms. These axioms can be used to derive all of the properties of P, the *probability function*. (These axioms are also known as Kolmogorov's axioms.)

For a finite sample space S,

Axiom A1: If A is any event defined over S, then $P(A) \geq 0$.
Axiom A2: $P(S) = 1$.
Axiom A3: If A and B are any two mutually exclusive events defined over S, then

$$P(A \cup B) = P(A) + P(B).$$

In the case that S consists of infinitely many elements, then the additional axiom needed to completely describe P is

Axiom A4: If $A_1, A_2, \ldots$ are events defined over S, and if $A_i \cap A_j = \emptyset$ for each i distinct from j, then

$$P\left(\bigcup_{i=1}^{n} A_i \right) = \sum_{i=1}^{n} P(A_i).$$

Exercise Set 43

Prove the following theorems for a sample space S:

1. **THEOREM 47:** For event A defined over S,

    $$P(A^C) = 1 - P(A).$$

2. **THEOREM 48:** $P(\emptyset) = 0$.

3. **THEOREM 49:** For events A and B defined over S, if $A \subset B$, then $P(A) \leq P(B)$

4. **THEOREM 50:** For any event A defined on S, $P(A) \le 1$.

5. **THEOREM 51:** For events A and B defined over S,

$$P(A \cup B) = P(A) + P(B) - P(A \cap B).$$

6. **THEOREM 52:** For events A_1, A_2, ..., A_n defined over S, if $A_i \cap A_j = \varnothing$ for each i distinct from j, then

$$P\left(\bigcup_{i=1}^{n} A_i \right) = \sum_{i=1}^{n} P(A_i).$$

Discrete Probability Functions

Given that a sample space for an experiment has either a finite or a countably infinite number of outcomes, then any function P such that

1. $0 \le P(s)$ for each $s \in S$ and

2. $\displaystyle\sum_{\text{all } s \in S} P(s) = 1$

is said to be a *discrete probability function*. Thus, for any event A, it follows that

$$P(A) = \sum_{\text{all } s \in A} P(s).$$

One of the most important discrete probability functions is the *Poisson distribution function*. This Poisson function describes extraordinarily well many real-world phenomena. The sample space for this function is the set of all nonnegative integers. Probabilities assigned to the sample outcomes for this distribution are given by

$$P(s) == \frac{e^{-\lambda} \lambda^s}{s!}, \text{ where } s = 0, 1, 2, \dots \text{ and}$$

λ is a constant based on the average number of outcomes per unit time or region. It is of interest to note that

$$\sum_{\text{all } s} P(s) = \sum_{s=0}^{\infty} \frac{e^{-\lambda} \lambda^s}{s!} = e^{-\lambda} \sum_{s=0}^{\infty} \frac{\lambda^s}{s!} = e^{-\lambda} e^{\lambda} = 1.$$

Exercise Set 44

1. A fair coin is tossed until a head comes up for the first time. What are that chances of this happening on an even-numbered toss?

2. Use the Poisson distribution to solve the following problem: in a laboratory experiment the λ associated with the average number of particles passing through a counter in 1 millisecond is $\lambda = 4$. What is the probability that that $s = 6$ particles enter the counter in 1 millisecond?

3. Research the Poisson distribution function online or in textbooks, and present an application of the function to the class.

Continuous Probability Functions

Just as we started with discrete numbers and found our way to the complete or continuous set of real numbers earlier in the text, we will now make the leap from discrete probability functions to *continuous probability functions*.

We label S as a *continuous* sample space if it contains a bounded or unbounded interval of real numbers. We associate with this S a *continuous probability function*, f, such that if A is an event defined over S, then f is a real-valued function having the property that

$$P(A) = \int_A f(x)dx.$$

Analogous to the previously defined discrete functions, the continuous probability function f must satisfy these two conditions:

1. $0 \leq f(x)$ *for each* $x \in S$ and

2. $\int_s f(x)dx = 1.$

At this point, we caution you: be ever cognizant that the distinction between discrete and continuous probability functions needs to clear in your mind. In the discrete case, $P(s)$ is the probability that an experimental outcome will be s. On the other hand, in the continuous case, $f(x)$ does not represent the probability of the experimental outcome being x. Rather, f describes the outcomes of an experiment in the sense that the probability of any event A is the integral of $f(x)$ over all the points of A.

An example of a continuous probability function is

$$f(x) = \frac{1}{\lambda}e^{-\lambda x}, \qquad \text{where } x > 0 \text{ and } \lambda \text{ is a positive constant.} \qquad (1)$$

This function is used in physics to describe probabilistically the distance X that a molecule travels before colliding with another molecule.

Exercise Set 45

Physicists denote the average distance between molecular collisions as

$$\mu = \int_0^\infty x f(x)\, dx.$$

This is called the *mean free path*. Use equation 1 to find the probability that the distance a molecule travels between consecutive collisions is less than half its mean free path. That is, integrate μ over the interval $[0, \dfrac{\lambda}{2}]$ using equation 1 as $f(x)$.

A Special Continuous Probability Distribution Function

By far, the most important and well known continuous probability distribution function is the bell-shaped *normal distribution function*, also known as the *Gaussian distribution function*. The sample space for this function is the entire real line. This probability distribution depends upon μ and σ, its mean and standard deviation. This continuous probability function is given by

$$f(x) = \frac{1}{\sqrt{2\pi}\sigma} e^{-1/2[(x-\mu)/\sigma]^2}, \qquad \text{where } -\infty < x < \infty, -\infty < \mu < \infty, \text{ and } \sigma > 0.$$

Exercise Set 46

1. Show the following:

$$1 = \frac{1}{\sqrt{2\pi}} \int_{-\infty}^{\infty} e^{-\frac{1}{2}x^2}\, dx$$

2. Use some calculus on the general formula for the standard normal density curve to find the points of inflection for the curve. Assume that $\mu = 0$ and $\sigma = 1$ in the formula.

$$Y = \frac{1}{\sqrt{2\pi\sigma^2}} e^{\frac{-(x-\mu)^2}{2\sigma^2}}$$

3. Use the fact that for continuous situations $E(x) = \int_{-\infty}^{\infty} x f(x)\, dx$ to show that for the standard normal density curve $E(x) = \mu$. (Hint: to evaluate the integral, let $z = \dfrac{x-\mu}{\sigma}$ which implies $dx = \sigma\, dz$.) Note that this process can also be used to show that $E[(X-\mu)^2] = \sigma^2$.

9. Another Algebraic Structure—Group

Recall, we chose as our starting point for consideration an *algebraic* structure known as a field. Thus, it is fitting that we close this text with consideration of another type of *algebraic* structure. The word "algebraic" refers to the fact that we can carry out, using the 11 axioms (A1-A5, M_1-M_5, D), calculations that are parallel to what we do when we do basic algebra; such as solving linear equations. Note that the set $\mathbb{Z}$ of integers is an interesting set that comes close to being a field. That is, $\mathbb{Z}$ satisfies 10 of the eleven axioms (which one is missing?).

Mathematicians have considered and studied sets of objects that satisfy some but not all of the 11 axioms, even considering only half of an axiom (For example; assume A1-A5, M_1, M_3, M_4, M_5, D*: $a(b + c) = ab + ac$). In the research literature one can find articles that study algebraic structures with names such *as ring, loop, semigroup, near-ring, nonassociative ring, quasigroup, group.*

Exercise Set 47

1. Verify that $\mathbb{Z}$ satisfies 10 of the eleven axioms.

2. Which axiom fails to hold in $\mathbb{Z}$ and why (i.e. provide an example)?

Observe that one cannot arbitrarily choose a set of axioms and expect to have a corresponding algebraic structure. For example, axiom M3 is meaningless without axiom M1; similarly, one needs A1 to have A4.

Interesting algebraic structures occur also without assuming the existence of two operations. Thus one can work with a subset of axioms A1-A5 or Axioms M1-M5 (observe that one must modify M5 dropping the restriction that $x \neq 0$, since 0 is a property associated with A1-A5). An algebraic system involving only one operation that has played a major role in the development of mathematical thought is that of a *group*. This type of structure occurs naturally within fields in two ways: One can take the elements of the field, use axioms A1-A5 and ignore the multiplication; or one can take the *nonzero* elements of the field, use axioms M_1-M_5 and ignore the addition.

For our purposes we will base our discussion on the model obtained by using the second example. We will, in fact, work with only 4 of the axioms, since interesting examples arise from the fields of Topology, Combinatorial Analysis, and Geometry. We choose a geometric example.

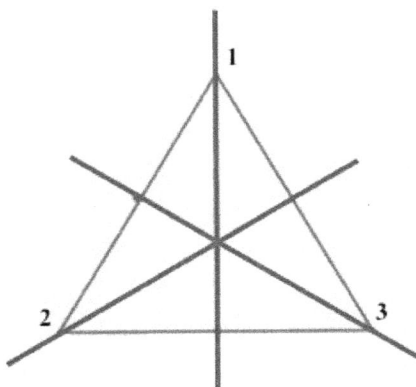

Figure 6

We start with a planar equilateral triangle whose center is located at the origin; label the vertices 1, 2, 3 according to the Figure 6. The triangle can be rotated counter-clockwise through angles of 120°, 240°, and 360° to bring the triangle back into a similar position. We can keep track of the rotations by seeing what happens to vertices. This permits us to think of a rotation as a function from the set {1, 2, 3} to the set {1, 2, 3}. Under a counter-clockwise rotation of 120°, vertex 1 moves to vertex 2, vertex 2 to vertex 3, and vertex 3 to vertex 1.

This rotation can be represented by the function P, when $P(1) = 2$, $P(2) = 3$, and $P(3) = 1$. Rotation of 240° can be represented by the function q, when $q(1) = 3$, $q(2) = 1$, and $q(3) = 2$. A rotation of 360° is the same as a rotation of 0°. This rotation is represented by the function I, where $I(1) = 1$, $I(2) = 2$, and $I(3) = 3$. A convenient way to represent these three functions follows:

$$P = \begin{pmatrix} 1\,2\,3 \\ 2\,3\,1 \end{pmatrix}, \; q = \begin{pmatrix} 1\,2\,3 \\ 3\,1\,2 \end{pmatrix}, \; I = \begin{pmatrix} 1\,2\,3 \\ 1\,2\,3 \end{pmatrix}$$

Note that the second row represents the corresponding value of the function evaluated at each term in the first row.

The triangle also has 3 axes of symmetry; these are lines passing through each vertex and the center. Reflecting the triangle through each axis of symmetry yields three additional functions:

$$a = \begin{pmatrix} 1\,2\,3 \\ 1\,3\,2 \end{pmatrix},\ b = \begin{pmatrix} 1\,2\,3 \\ 3\,2\,1 \end{pmatrix},\ c = \begin{pmatrix} 1\,2\,3 \\ 2\,1\,3 \end{pmatrix}$$

You should check these results with the help of Figure 6.

If we collected these six functions into a set $M = \{I, P, q, a, b, c\}$, there is a natural way in which we can combine any two functions from a group G and still obtain a function belonging to M, namely the *composition* of functions.

We can then consider the set M with composition as an operation on the elements of M and obtain an example of a *group*. The formal definition follows.

> **Definition.** A group G is a set with an operation such that the following four conditions hold:
> G1: For all $x, y \in G, x * y \in G$ (closure)
> G2: For all $x, y, z \in G$,
> $(x \cdot y) \cdot z = x \cdot (y \cdot z)$
>
> G3: There exists an element $i \in G$ such that for all $x \in G$,
> $ix = x$ and $xi = x$
>
> G4: For each $x \in G$, there exists $y \in G$ such that $xy = 1$ and $yx = i$.

Exercise Set 48

1. Verify that the set M, together with the operation of the composition of functions

   (denoted by $\circ$) satisfies G_1 through G_4.

2. Find functions $x, y \in M$ such that $x \circ y \neq y \circ x$.

Exercise Set 49 (See chapter 1 as a refresher, if needed)

Show that:

1. The element i identified in G_3 is unique.

2. The element y identified in G_4 is unique.

3. An alternate way to consider the group M is to consider the set of functions from the set $\{1, 2, 3\}$ to $\{1, 2, 3\}$ which are both 1-1 and onto. It is easy to check that there are exactly six such functions. (Note that being 1-1 ensures that a function is also onto and vice-versa).

4. Let K be a finite set and let $f : K \rightarrow K$ be a function from K to K.

    a. Show that if f is 1-1, then f is onto.

    b. Show that if f is onto then f is 1-1.

A Permutation Group on S

Note that the group consisting of the six functions from $\{1, 2, 3\}$ to $\{1, 2, 3\}$ that are one-to-one and onto, with the operation of composition, is commonly denoted by S_3 and is referred to as

- The *symmetric group* on two symbols

- The *symmetry group of an equilateral triangle*

- The *permutation group on the set* $\{1, 2, 3\}$

> **Definition:** Let S be a set. A permutation of S is a function from S to S, which is both one-to-one and onto.

For each integer $n \geq 3$, we can consider the symmetries of any regular n-sided polygon centered at the origin and represent them by the group of all permutations of the set $\{1, 2, \ldots, n\}$ using composition of functions as the operation. This group is denoted by S_n.

Exercise Set 50

1. Verify that S_n is a group.

   If $3 \leq k < n$, there is a natural way to consider S_n as being contained in S_n. (Hint: Consider those functions f, such that $f(i) = i$ for all $k < i \leq n$.)

   Note: The set of all one-to-one and onto functions from $\mathbb{Z}$ to $\mathbb{Z}$ is a group S_n; thus, the same can be said for any set S_n.

2. Provide an example of a group existing within another group relative to the same operation. We refer to smaller groups as a *subgroup* of the bigger group.

3. Prove the following: A subset H of group G is a subgroup if and only if the following two conditions are satisfied:

   a. If $a, b \in H$ then $ab \in H$.

   b. If $a \in H$, then $a^{-1} \in H$

Last, we provide some parting food for thought. You may have seen or worked with some of these theorems in the past. The hope is that a bigger-picture view of this theory has been afforded to you as a result of your work with this course.

Given a subgroup H of a group G, a related subset of H is what is known as a *right coset*, obtained by taking any element $a \in G$ and forming the set $Ha = \{xa \mid x \in H\}$.

Theorem 53

Let H be a subgroup of a finite group G. Then

a. $Hg \in H$ if and only if $g \in H$.

b. $Ha = Hb$ if and only if $a \in H$.

c. If $Ha \cap Hb \neq \emptyset$, then $Ha = Hb$.

For any $a \in G$, the number of elements in Ha is the same as the number of elements in H. When G is a finite group, we write $o(G)$ to indicate the number of elements in G and refer to it as the *order of* G. We can now make one application to number theory. The notation used is that found in chapter seven.

Theorem 54

In the field $\mathbb{Z}_p$, an element $a \in \mathbb{Z}_p$ is a *quadratic residue* modulo p, if the congruence $x^2 \bmod p$ $x^2 \equiv a \bmod p$ is solvable.

Let p be a prime number. The set of quadratic residues mod p is a subgroup of the group of $\mathbb{Z}_p{}^* = $ *nonzero elements of* $\mathbb{Z}_p$.

Exercise Set 51

Determine the order $\mathbb{Z}_p{}^*$ and the subgroup of quadratic residues. The subgroup of quadratic residues is denoted by $g\left(\mathbb{Z}_p{}^*\right)$.

Theorem 55

a. If $a, b \in g\left(\mathbb{Z}_p{}^*\right)$ then $ab \in g\left(\mathbb{Z}_p{}^*\right)$.

b. If $a \in g\left(\mathbb{Z}_p{}^*\right)$ and $b \notin g\left(\mathbb{Z}_p{}^*\right)$ then $ab \notin g\left(\mathbb{Z}_p{}^*\right)$.

c. If $a \notin g\left(\mathbb{Z}_p{}^*\right)$ and $b \notin g\left(\mathbb{Z}_p{}^*\right)$ then $ab \notin g\left(\mathbb{Z}_p{}^*\right)$.

Conclusion

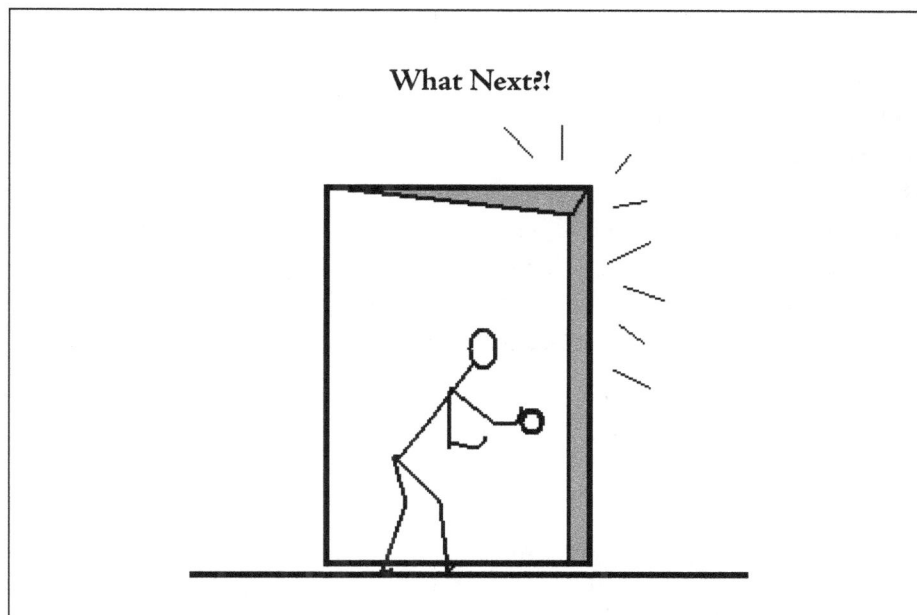

What Next?!

Concluding Remarks from the Authors

We hope that your journey through this text has provided you with some insight into the existence of and relationship between numbers both on the number line and in the plane. While it is easy to take for granted the fact that different kinds of "numbers" and fields exist, the justification of the existence of these entities involves an axiomatic approach that unites many different branches of mathematics. Topics that you have explored as a learner/reader of this text were taken from abstract algebra, analysis, geometry, linear algebra, and set theory.

It is also our hope that you have gained an appreciation of the "connectedness" of the topics presented in this book in relation to the different mathematics courses that you have completed thus far in your studies. The topics presented here just brush the surface of the mathematical connections that could be explored relating to "numbers" and the properties and connections associated with the construction and use of numbers. You are encouraged to view these topics as a springboard for further discovery and exploration along these lines. Other related topics that you might pursue are complex analysis, coding theory, and various connections in applied mathematics.

Last, it is our belief that your journey through the material of this text has helped prepare you for future graduate studies in mathematics. After all, often at the heart of graduate work in mathematics is the creative, in-depth search for the "connectedness" of seemingly disparate topics in mathematics. Additionally, for those who are interested in teaching mathematics at various levels, a journey through this book has provided a background and ability for you to convey the importance of the foundation of number systems and fields in mathematics. The authors hope that the completion of the material of this text by students provides not an end but a beginning to a new and more refined view of mathematics in light of graduate studies or dissemination of the importance of the foundations inherent to the study of mathematics.

Index